Quality Management System

A Planning and Auditing Guide

Walter Willborn, Ph.D.
Faculty of Management
University of Manitoba, Canada

Industrial Press Inc.

Library of Congress Cataloging-in-Publication Data

Willborn, Walter W. O.
  Quality management system.

  Includes bibliographies and index.
  1. Quality assurance—Management. I. Title.
TS156.6.W54   1988   658.5'62   88-35763
ISBN 0-8311-3013-X

INDUSTRIAL PRESS INC.
200 Madison Avenue
New York, New York 10016-4078

QUALITY MANAGEMENT SYSTEM:
A Planning and Auditing Guide—First Edition

Copyright © 1989, by the Industrial Press, Inc., New York, NY. Printed in the United States of America. All rights reserved. This book, or parts thereof, may not be reproduced in any form without the permission of the publishers.

Composition by Edwards Brothers, Inc., Ann Arbor MI.
Printing and binding by Quinn Woodbine, Woodbine, NJ.

2   4   6   7   5   3

Preface

"Quality is never an accident, it is always the result of intelligent effort," wrote John Ruskin, the great Victorian art critic. This is true in business and industry and in private or public organizations with regard to products and services. Specially designed and implemented quality management systems direct all decisions and actions toward well-defined quality goals. Managers take the responsibility for this system, yet they often lack sufficient technical expertise. It must be remembered that "quality starts at the top"!

Purpose of this *Guide*

The purpose of this *Guide* is to

* inform the reader about concepts and methods of quality assurance,
* assist in planning, implementing, and maintaining quality assurance procedures,
* serve as a general reference source,
* save money and enhance effectiveness of quality assurance,
* complement textbooks, standards, and technical aids.

This *Guide* is not

* to be used for establishing quality assurance procedures without adaptation to prevailing conditions,

* to teach quality assurance as a textbook would,
* a handbook.

Approach

The reader is assumed to have basic knowledge in management. We shall raise the reader's awareness of quality assurance as an important responsibility in all business establishments and in private or public organizations. After clarification of the major benefits derived from quality assurance, the reader will be motivated to study the need for and applications of quality assurance in his or her own company, organizational unit, or workplace. Actual planning for quality assurance requires certain preconditions that we shall explain.

In its main sections this *Guide* presents prototype procedures for quality assurance; these procedures range from product or service design, supplies, production, operations, distribution to customer services. Each procedure has the same format, which explains the purpose, policy, definitions, references to standards, application, steps for implementation, and forms/technical aids. A general explanatory note expounds further aspects of the procedure and the activity.

Application

This *Guide* is written for small and large organizations, but some procedures especially designed for small enterprises will be so indicated. This *Guide* does not pretend to be valid and applicable in every situation where quality assurance procedures are to be established. Sometimes the existing practices only need to be documented and declared as binding procedures. In other cases, managerial capacities need to be further developed before quality assurance can be properly planned and introduced. But this *Guide* will always be helpful.

Managers will benefit from this *Guide* because quality assurance is a new field, for most of them. In the past, engineers and quality experts had the responsibility for planning and controlling quality. This situation has changed, and managers must also apply their expertise in management to quality assurance.

Quality assurance specialists can use this *Guide* in many ways, such as a reference or training aid, or initiating quality assurance activities and drafting policies and procedures. Students and their instructors will find this *Guide* a useful enrichment and information source about quality assurance practices.

References

Several standards describe elements of quality assurance systems. The reader must know of these standards and will be informed accordingly. These reference documents mainly describe stipulations, but do not provide prototype procedures as they are found in our *Guide*. Consequently, reference standards and our *Guide* complement each other from the viewpoint of the practitioner.

Topical Layout

As shown in the Table of Contents, we introduce the reader to quality assurance with direct reference to the task at hand—the planning of a quality assurance system. Persons less familiar with the subject matter will be adequately informed. The procedures that follow serve as a first draft and can be adapted to the company. The approach to procedure writing and adoption of our prototype procedures will be explained to the reader. Suitable procedures can be selected and adopted, others will have to be omitted in the company's quality assurance system. Our *Guide* offers this flexibility.

Planning, implementing the plan, and controlling and auditing the execution are logical management steps. We have arranged our sections accordingly. In each phase of the system development, however, we present respective procedures, which throughout are written in the same format and layout for convenient application.

Acknowledgments

I have worked with many practitioners over the years and have had the benefit of their guidance and encouragement. *Quality Management System: A Planning and Auditing Guide* greatly reflects their contribution. My colleagues and students have also helped me in many ways, which I fully acknowledge.

Contents

Preface		v
1	**Preparation for Planning Quality Assurance**	3
	1.1 What is a "Quality Management System"?	3
	1.2 Developments in Quality Assurance	4
	1.3 Questions Frequently Asked about Quality Assurance	5
	1.4 Quality Assurance Tasks	14
	1.5 Quality Assurance System Standards	18
	1.6 Quality Cost Accounting	21
2	**Designing the Quality Assurance System**	25
	2.1 An Action Plan	25
	2.2 Project Planning and Control	31
	2.3 Quality Improvement Projects	36
	2.4 Procedure Writing Projects	40
3	**Quality Assurance Procedures: Subsystems**	45
	3.1 Design Assurance and Contract Review	45
	3.2 Quality Plan or Inspection Plan	57
	3.3 Supply Assurance	61
	3.4 Production Assurance	74
	3.5 Product Performance and Customer Service Assurance	93
	3.6 Quality Management Information System	100

3.7	Statistical Quality Planning and Control	107
3.8	Software Quality Control	115
4	**Implementing the Quality Management System**	**123**
5	**Auditing the Quality Assurance System**	**139**
	5.1 Audit Checklists	146
6	**Cases**	**161**
	6.1 Hospital Quality Assurance System	161
	6.2 A Small Manufacturer's Quality Assurance System	173
	6.3 Quality Assurance in a Research Establishment	177

Appendix A: Glossary — 191

Appendix B: Hospital Quality Assurance Manual (Outline) — 197

Appendix C: Quality Manual for a Small Manufacturer (Outline) — 203

Index — 213

Quality Management System

1
Preparation for Planning Quality Assurance

Before we commence planning and designing quality assurance procedures, an overview of this subject is necessary. I will discuss the meaning of "quality management system," "quality assurance," and other key terms used in this Guide. I shall also phrase some questions that are often asked, and also briefly discuss the general features of our action plan.

1.1 What is a "Quality Management System"?

Our "management system" entails interrelated procedures directed toward accomplishing predetermined goals. The purpose and goal of a quality management system are to ensure and attain the desired and specified quality of products and services.

The "quality of a product or service" is the desirable composite of all features and attributes that satisfies the needs and expectations of the customer. Confidence in the delivery of this quality rests on the reputation and "quality image" of the supplier and his or her specific quality assurance activities.

A "quality assurance system" embraces all individual procedures and directs all work and workmanship toward the attainment of quality and customer satisfaction. A quality assurance system has to be designed, implemented, and maintained carefully. Many factors, such as price, cost, and severity of defects, influence the system design. In its final form in a company this quality assurance system is called a "quality assurance program."

1.2 Developments in Quality Assurance

Customers have always insisted upon satisfactory quality. However, we know that products and services have changed. With advanced technology, the principle of "buyer beware" is not fair or acceptable when the customer cannot determine the quality of the product adequately. In addition, "just in time" delivery schedules do not allow for the return of unsatisfactory supplies. Therefore, industrial and private customers must rely on the assurances given by the supplier.

The following list contains some of the developments in quality assurance.

* "Zero defects" is the goal.
* Major customers insist on documented quality assurance systems, and national and international standards stipulate details of such systems.
* Strong competition in the marketplace requires quality assurance as a special service to customers.
* Auditing of quality assurance systems without the system's procedures being fully documented is impossible.
* Senior management must be aware of the need for and be committed to formal quality assurance.
* Modern assurance science provides quality control techniques that add competitive strength.
* Individual quality control measures and activities need to be organized, interrelated, and streamlined through an overriding quality assurance program.
* Costs of failures and defects have risen, making their prevention through quality control more cost-effective.
* Quality assurance is widely delegated to staff and integrated with other responsibilities.
* Employers and employees recognize that quality assurance contributes to company growth and job security.

* Workers are motivated to participate in quality assurance in various ways, improving their own workplace.

* Quality assurance has become an important subject in management education.

1.3 Questions Frequently Asked about Quality Assurance

Why should a company have a quality assurance program?

The benefits of a quality assurance program in terms of higher profit, increased productivity, larger market share, job security, "quality image," customer loyalty, staff cooperation, better workmanship, better utilization of economical and technical resources can be considerable. These benefits exceed their costs when quality assurance is planned, implemented, and maintained properly. In a competitive environment with rapid technological progress the quality of products and services normally can be improved through effective quality assurance measures.

If a company has every indication that the quality of its products and services fully satisfies its customers, why does it still need a quality assurance program?

Assuming that customers are fully satisfied sounds complacent. What might be true today will not necessarily hold true tomorrow. In dynamic and effective management we anticipate new customer quality demands and expectations. The quality assurance program serves as a vehicle for monitoring opportunities for quality improvements and for opening new avenues to serve old and new customers better.

Does not any well-managed company have quality assurance?

Yes, this is always true, when a company has established itself firmly in the marketplace and enjoys a positive quality image. We distinguish, however, informal quality assurance practices from formal and documented procedures. When a company grows, rules and directives become necessary for consistent cooperation and predictive decision making. For external audits or liability claims, the documented quality assurance program becomes essential as objective evidence.

Should I adopt the program of a competitor, or do I need to design it from scratch?

Reinventing the wheel is not necessary in quality assurance. One should and can learn from a strong competitor. The access to information is of course limited. Adopting a quality assurance program has major drawbacks because each company faces unique situations. Consequently, a company's management must design its own suitable quality assurance program. In this design process, careful preparation, information gathering, and auditing help management to establish the "right" and most cost-effective program.

Does a small company need special quality assurance and a formalized program?

The need for a quality assurance program arises more from the kind of products and services and the inherent effects of defects and failures, than from the size of the company. When the repercussions from defects are serious, a comprehensive quality assurance program starting with design assurance is warranted. Small enterprises with less risky products and services and a well-established clientele might not require a formalized program. But what happens when major customers require one? Should the company then forego the opportunity for growth?

What kind of help can I get for designing and implementing the program?

Designing your own quality assurance program might not be difficult.—We hope that this Guide will help sufficiently.—Managers, knowing the quality expectations of their major customers, need to translate these expectations into explicit assurance procedures down the line. Supervisors, suppliers, engineers, and accountants as well as quality assurance specialists or even outside consultants can be of help. When specialists are required, associations such as the American Society for Quality Control, Milwaukee, WI, help through advertisements in their journals. There are many technical support services for any kind of assurance problem. Standards and guides are the more important and less costly. We shall refer to them frequently.

What useful standards and guides do you suggest as a technical support?

Documents that describe major elements and approaches in quality assurance and have passed rigorous reviews before publication differ somewhat. Some have been prepared for procurement and contractual purposes; others lend themselves as information source for self-designed programs and are phrased in less legalized language. In our Guide we list the major standards and guides and also help in selecting the appropriate document.

What other books useful for the designing of a quality assurance program do you recommend?

There are many useful and interesting books on quality assurance. As opposed to standards and guides, books outline quality assurance concepts, practices, and methods in more detail. Most of them are listed in the annual catalog of the American Society for Quality Control, 310 West Wisconsin Avenue, Milwaukee, WI 53203 (call toll-free 800-952-6587). Under "Managing Quality" one finds books by major authors, such as Crosby, Deming, Feigenbaum, and Juran. All books have brief accompanying comments in this catalog.

Are courses and seminars offered that can be helpful?

Yes, there are some good ones and some not so good ones. The American Society for Quality Control provides many seminars, and you should call for a listing (see preceding question). Local sections of ASQC and other associations also arrange courses and seminars for practitioners; normally these courses and seminars are designed for supervisors and technicians. A local college or university might have suitable course offerings, because the demand for higher-level studies is rising. Seminars offered by private persons and institutions should be checked and their quality verified.

What company functions should be included in the program?

That depends on the scope of the program and possibly the standard invoked. For simple mature products, end-item inspection might suffice. In other situations production processes and supplies are included. The most comprehensive program accompanies all "life-phases" of a product or service, from the inception, design, procurement, and production operations to distribution/installation and customer services.

Functions in a company, such as marketing, accounting, staffing, and public relations, become involved in quality assurance in various ways. Under so-called "total quality control," quality assurance responsibilities become written into every position. The principle "quality starts at the top" means that senior management plays an important and visible role. "Quality is everyone's concern" signals the permeation of quality assurance participation across all function and job-level boundaries.

Should designers be included?

It has become more cost-effective and necessary to "build quality into the product" during the design process rather than to "inspect it in" at a later stage. Designers are usually highly qualified and motivated creative persons who feel that quality assurance procedures hinder more than promote the design process. Designers should be included in the quality assurance effort in all phases of the system design, implementation, and maintenance. With their input and the help of the quality assurance specialist, better review schemes and procedures are developed and become accepted. The quality specialist represents the customer in the design process.

Is quality assurance of production processes and operations, perhaps just an end-item inspection, sufficient?

The scope of a quality assurance program always depends on both the impact and the cost of failures and defects. Costs rise when the detection of a major defect is delayed during the production and distribution processes. When the customer discovers the defect, it is even more costly. Nevertheless, one should not generalize that the earliest inspection and detection are best. A general decision rule suggests that the defect/failure costs weighted by the probability of occurrence should be less than the prevention and inspection costs per unit. We shall deal with quality cost accounting procedures in more detail.

Who should design the quality assurance program?

Delegating the designing of a quality assurance program to one specialist, or even a nonspecialist, is neither advisable nor usual. Quality assurance must involve many persons inside and outside the company. The earlier this is done in the life of a quality assurance program, the better. The most appropriate arrangement is to form a committee or project team. The project manager does not necessarily need to be a specialist, particularly when the program is simple and our Guide is available to this person. A specialist might serve better as facilitator and coordinator in technical subprojects.

How should the designing and implementing of the program be organized?

Both tasks and activities have a particular goal and outcome; they therefore describe projects. Such a project consists of various interdependent activities that together can be seen as a network. The project manager has a particular mission and the authority to draw team members as they are needed from participating departments. The project planning and execution are best performed with special computer-based techniques, which allow flexible controls and adjustments.

In simple quality assurance programs and small business settings, the manager/owner or a staff member takes on the tasks, possibly being temporarily relieved of other duties.

What rule and principles in establishing a quality assurance program should be observed?

A participatory management style is best suited for this task, because quality assurance is basically a responsibility involved in every job, although often not written into the job description. Those who are to comply with the quality assurance procedures should be involved in their drafting. Once they receive sufficient training and indoctrination, procedure drafting can be delegated to other people entirely. Senior management and/or the quality assurance coordinator will review the drafts for consistency and general acceptability.

The quality assurance program should be reviewed and audited regularly in order to maintain its suitability and effectiveness. Benefits of the program should visibly and convincingly exceed the costs and general investment in time and energy.

The establishment of the program should be conceived and handled as a task primarily for management and only technical matters should be assigned to specialists.

What are criteria for a sound program?

From the management point of view cost effectiveness is most important. This requires quality cost accounting procedures along with other quality and quality assurance reporting. Other criteria are actual quality improvement, higher staff motivation, and tangible indications for customer satisfaction with products and services. Passing external audits is also a positive sign for a sound quality assurance program.

Recognized internal and/or external quality auditors and specialists must approve the program after examining it against valid program standards. The program could possibly be officially registered in conjunction with such audits.

Other indicators for a sound program are the acceptance by customers, cooperation from staff members, gradual decrease of the sum of failure/defect costs and prevention/appraisal costs ("quality cost"); positive developments in quality indices, such as less customer complaints, or fewer defects per period or unit.

Can one adopt or modify the quality manual of another company or a quality manual that is published in a book?

A quality manual describes the quality assurance program and includes all respective procedures. In it, the company interprets a standard into the company's own terms. It is not advisable to just copy a quality manual, because the situation for each company is unique and the underlying standards might differ. The imposition of a foreign quality manual and procedures can be counterproductive. Companies must change individual procedures and their manual frequently in order to keep both of them valid and acceptable.

Quality manuals of similar companies and those published can still provide useful guidance. The prototype procedures in this Guide should serve this purpose.

What is a quality manual and what is its use?

The quality manual documents the quality assurance policy of a company, or one of its independently managed branches or plants. Moreover, the respective organization, all the individual quality assurance procedures, and documentation are outlined in it. Standards for more comprehensive programs stipulate the major part of these manuals, but leave the format to the company.

A quality manual informs management, employees, suppliers, regulatory bodies, and, in particular, major customers and their auditors about the program. The quality, validity, and reliability of the manual hinges on the actual implementation and compliance by those concerned in the company. Dynamic quality assurance and quality improvement will be reflected in adjustments to the manual.

Once the quality assurance program is designed and documented, what are the rules and principles for implementation?

Implementation means that those personnel addressed by the procedures learn to comply with them and observe proper quality assurance. Moreover, the affected personnel can be expected to induce procedural changes when these are deemed necessary, and, possibly after some additional training, should participate in the writing of procedures. This means that implementation actually should start during the design phase, where the specialist and the project team supervise procedure writing for consistency with policies and proper methodology.

How can one keep the Program effective and useful?

First, design and implementation must be sound, otherwise the program is not effective and useful from the start. Over time, any individual procedure or the whole program becomes obsolete as internal and external developments occur. Those personnel and their supervisors who are to comply with the procedures should initiate updates and adjustments to the program. Personnel in charge of the program, the quality assurance manager, or the coordinator frequently observe, review, and audit compliance and program effectiveness.

Audits have been developed as independent and formal examination devices, with reports going to senior management and also to those responsible for the audited area. Standards govern audits and guide auditors.

In auditing the quality assurance program, including suppliers' programs, what are technical supports? What are the major rules and principles?

Audit standards and guidelines have been published and are listed and described in this Guide. The guideline for auditing quality systems (ANSI/ASQC Q1-1986) lists some useful books for auditing and includes a flowchart of auditing steps. Some standardized checklists have been published; and institutes that register quality programs provide audit services.

Audits are formal and systematic examinations of quality assurance with regard to compliance with applicable quality assurance program standards and to general program effectiveness. The quality assurance program to be audited must be documented and established. The auditor must be independent and qualified; the American Society for Quality Control (ASQC) offers an auditor certification program. The audit is planned carefully, with the object and objectives clearly determined. Auditees should cooperate with auditors, and are informed about the auditor's observations and reports. Management initiating the audit receives the audit report and decides on corrective action, possibly a follow-up audit.

Should the regular finance or internal auditors audit the quality assurance program?

Why not, if they have adequate knowledge about the particulars of a quality assurance program. However, even though audit principles and approaches are universal, quality assurance auditing is a recently developed and rapidly expanding field, therefore specially trained and experienced quality auditors are preferred, particularly for external audits of suppliers. Joint audit teams and audits have also been proposed and arranged in some companies.

QUALITY MANAGEMENT SYSTEM

How can unnecessary paperwork and "red tape" in the quality assurance program be avoided? Is there a time when the program should be terminated?

One of our major goals of this Guide is to help the reader avoid establishing a bureaucratic "Frankenstein monster." Every bit of data compiled, analyzed, recorded, and communicated must serve some known and practical purpose. Procedures that are redundant should be eliminated; those procedures that are missing should be added.

How does the person in charge make these decisions? He or she evaluates the situation through frequent critical observations, reviews, and audits.

It is difficult for a company to survive in the marketplace against strong competition without a quality assurance program of some sort. The problem management faces is to implement the most suitable and effective program. As conditions and situations change—for instance, the company grows from small to large, new products are added, new personnel are hired, or new processes and technology are installed—the old quality assurance program must be replaced by a more suitable one. Simple modifications often do not suffice.

Should one prepare special quality assurance programs for different customers or for different products/services/contracts?

This is a frequently asked question, and has no easy answer. One major customer and its auditor may accept the quality assurance program, another customer wants alterations to it. Perhaps the second customer refers to different standard or interprets the same standards differently, or the requirements of the products differ greatly.

As a general rule a basic quality assurance program with some built-in flexibility should be established, possibly for the minimum rather than the maximum demands. The product-related quality plans then can cater to special needs of the product and its customer. Additional and more specific quality plans describe design, supply, and production requirements, and include test and inspection plans, for one particular product.

As a last resort negotiation about conflicting quality assurance requirements might resolve the matter. Use of acknowledged program standards limits, if not prevents, conflicts.

In what way are the prototype procedures published in this Guide useful? What are their pitfalls?

The reader will have to determine this. We hope, however, that our listings allow careful selection and adaptation. Similar to quality assurance program standards, our procedures are generic and generally applicable. They are designed to be modified in the light of what is useful and required in each situation they are applied. Uncritical copying and adoption of the procedures is the major pitfall; false or invalid procedures are harmful. In addition, imposing procedures rather than developing them in cooperation with all personnel leads to difficulties and resistance.

As with any prototype, it is something new and still unproven to a certain extent. Writing the correct procedures is a continuing task in the life of a company and its management. Comments on our prototype procedures and suggestions for improvement will be welcomed.

1.4 Quality Assurance Tasks

The goal of satisfying your customers with your company's product and service quality and quality assurance leads to numerous tasks and responsibilities. In the following subsections we list the major tasks of each organizational unit and level of management.

1.4.1 TOP MANAGEMENT

The responsibilities of top management in a quality assurance program are as follows:

* Setting explicit quality objectives.
* Formulating a quality assurance policy.
* Initiating and supervising a quality assurance program.
* Conducting audits of the compliance with standards and effectiveness of the quality assurance program.

QUALITY MANAGEMENT SYSTEM 15

* Providing resources to implement the quality assurance program.

1.4.2 MARKETING

The responsibilities of the marketing department in a quality assurance program are as follows:

* Conducting quality-related market research and analysis.
* Determining product specifications and associated services.
* Establishing price and quality relationships.
* Preparing advertisements and other documents emphasizing the quality aspects of the company's products and services.
* Surveying the quality and quality assurance of competitors.
* Training sales personnel and distributors in all aspects of the quality assurance program.
* Providing customer service.

1.4.3 DESIGN ENGINEERING

The responsibilities of design engineers in a quality assurance program are as follows:

* Researching quality aspects and requirements for each of the company's products.
* Developing the technical details (quality characteristics) of products and services.
* Preparing quality specifications and resource requirements.
* Complying with legal and safety requirements.
* Standardizing product and parts design.
* Testing and verifying quality requirements.
* Documenting final design.
* Initiating design reviews.

1.4.4 PRODUCTION

The responsibilities of the production department in a quality assurance program are as follows:

* Clarifying quality aspects of design.
* Preparing the production plan and the associated inspection/test plan.
* Determining the requirements for and providing adequate technical and human resources.
* Setting workmanship standards.
* Determining the requirements for handling and storing facilities.
* Conducting test runs.
* Assessing and establishing the process capability.
* Instituting production controls and inspections.
* Analyzing nonconformances and correcting their causes.
* Ensuring proper shipping and transporting.

1.4.5 PROCUREMENT

The responsibilities of the procurement department in a quality assurance program are as follows:

* Clarifying requisitions.
* Selecting and negotiating with qualified suppliers.
* Determining quality assurance requirements and conducting supplier surveys and audits.
* Verifying the quality of deliveries.
* Maintaining a product's quality during the transporting, storing, and handling of the product or contracting for such arrangements.
* Assessing and recording supplier performance.

1.4.6 HUMAN RESOURCES

The responsibilities of the human resources department in a quality assurance program are as follows:

* Determining staff requirements and qualifications.
* Preparing job descriptions including quality assurance responsibilities.
* Providing conditions conducive to good workmanship.
* Training, educating, and motivating the personnel to achieve the company's quality goal.
* Recognizing and rewarding outstanding performance.

1.4.7 ADMINISTRATION

The responsibilities of the administration in a quality assurance program are as follows:

* Establishing a quality management information system.
* Organizing the quality assurance function.
* Coordinating quality assurance and quality improvement projects.

1.4.8 ACCOUNTING AND FINANCE

The responsibilities of the accounting and finance department in a quality assurance program are as follows:

* Establishing a quality cost accounting system.
* Conducting quality benefit/cost and investment analyses.
* Preparing quality cost reports.

1.4.9 GENERAL SERVICES

The responsibilities of the general services department in a quality assurance program are as follows:

* Securing the safety of facilities and workplaces.
* Assessing and controlling environmental conditions.

1.4.10 QUALITY ASSURANCE

The responsibilities of the quality assurance department in a quality assurance program are as follows:

* Preparing the quality assurance program and quality manual.
* Initiating and coordinating quality improvement projects.
* Preparing inspection and test plans.
* Verifying the process capability.
* Conducting workshops.
* Reviewing and auditing quality assurance.

Quality assurance activities are organized in the individual departments and are carried out for the various products, contracts, orders of the company. Procedures will be prepared for the design, procurement, preparation, production, and delivery phases, in general, and again for each item and unit of production. Quality assurance is realized in the individual production systems and projects. Just assigning quality assurance responsibilities to departments does not suffice. Standards for quality assurance programs imply a dynamic, goal-oriented system rather than a hierarchical organization and assignment of jobs. What is required are actions, and the following chapters describe a quality assurance action plan after quality assurance standards have been introduced.

1.5 Quality Assurance System Standards

Standards that describe the major features and requirements of quality assurance systems serve as the most useful information source for managers when they establish their quality assurance program.

The table of contents as shown in Table 1-1 is typical for all standards of this kind.

Standards must be flexible. Some companies require only a simple system with only end-item inspection; other companies need a more comprehensive system that includes control of production and

Table 1-1. Typical Quality Assurance System Standard Contents

1. Scope
2. Definitions
3. Requirements
   a. Quality assurance program
   b. Organization
   c. Audits
   d. Quality program documents
   e. Verification of quality
   f. System functions
      * Contract review
      * Design assurance
      * Document control
      * Measuring and testing equipment
      * Purchasing
      * Incoming inspection
      * In-process inspection
      * Final inspection
      * Inspection status
      * Identification and traceability
      * Handling and storing
      * Manufacturing and construction
      * Special processes
      * Preservation, packaging, and shipping
      * Quality records
      * Nonconformance
      * Customer-supplied items
      * Corrective action

even of design. One multitiered standard is the Z299 series of the Canadian Standards Association. One user of this series, Ontario Hydro, summarizes the four tiers in Table 1-2.

Guides, when compared to these standards, have a more explanatory content; they are not designed as mandatory procurement documents. In Table 1-3 we show the contents of "Quality Management and Quality Systems Elements—Guideline," International Standards Organization, ISO 9004, which is adopted in the United States, in Canada, and in many other countries. The American Society for Quality Control sells this document; the U.S. version is ANSI/ASQC Q94-1987.

Table 1-2. Main Features of Quality Assurance Standard CSA Z299

Z299.4	Z299.3	Z299.2	Z299.1
			Category 1[a]
			Preventing
			★ Management review ★ Tender review ★ Design planning ★ Process review ★ Internal audit
		Category 2[a]	
		Reacting	
		★ Design verification ★ Production planning ★ Program procedures ★ Corrective action	
	Category 3[a]		
	Verifying		
	★ Manual ★ Competent personnel ★ Inspection plan ★ Program descriptions ★ Documentation ★ Procurement ★ Special processes ★ Statistical techniques ★ External audits		
Category 4[a]			
Sorting			
★ Management responsibilities ★ Contract review ★ Planned inspection ★ Calibration ★ Quality records ★ Disposition ★ Optional manual and inspection and test plan			

[a]Each category contains all of the features of all of the lower categories.

Table 1-3. Table of Contents of ISO Guidelines

1. Introduction
2. Scope and field of application
3. References
4. Definitions
5. Management responsibility
6. Quality systems principles
7. Economics—quality-related cost considerations
8. Quality in marketing
9. Quality in specification and design
10. Quality in procurement
11. Control of production
12. Product verification
13. Control of measuring and test equipment
14. Nonconformity
15. Corrective action
16. Handling and postproduction functions
17. Quality documentation and records
18. Personnel
19. Product safety and liability
20. Use of statistical methods

In the Table 1-4 we list the most widely adopted and generic documents.

For the selection of the most suitable document we recommend the approach of Table 1-5.

1.6 Quality Cost Accounting

Establishing a quality assurance system demands an investment of time and money. (Although it has been said that "quality is free," it can mean only that any company has dormant resources for quality improvement). More important, the quality assurance system must become a profit center. In order to build cost-effectiveness into the system design, managers need to know about quality costs, so that they can compile and analyze them. A special procedure will be presented subsequently in this Guide.

"Quality cost" is the sum of defect/failure prevention costs, appraisal (inspection) costs, and defect/failure costs (internal and external); they are accounted for in total business cost records and

Table 1-4. Standards for Quality Assurance Systems

International Standards Organization (ISO)

ISO/DIS 9000	Quality Management and Quality Assurance Standards & Guideline
ISO/DIS 9001/3	Quality Systems (three models)
ISO/DIS 9004	Guidelines

NATO Allied Quality Assurance Publications

AQAP-1/-4	Quality Control/Inspection Systems
AQAP-2/-5	Guides for System Evaluation

American National Standards[a]

ANSI/ASQC Q91-87	Quality Systems—Design/Development Production, Installation, Servicing
ANSI/ASQC Q92-87	Quality Systems—Production and Installation
ANSI/ASQC Q93-87	Quality Systems—Final Inspection/Tests
ANSI/ASQC Q94-87	Quality Management and Quality Systems Elements—Guidelines
MIL-Q-9858A	Quality Program Requirements
MIL-I-45208	Inspection System Requirements
ANSI/ASME NQA-1	Quality Assurance Program Requirements for Nuclear Facilities

Canadian National Standards

CSA Z299.1/.4	Quality Program Standards
CSA Z299.0	Guide for Selecting and Implementing the CSA Z299 Quality Programs

British Standards Institution

BS 5750	Quality Systems
BS 4891	Guide to Quality Assurance

[a]ANSI/ASQC Q90–Q94 are technically equivalent to the ISO 9000 Series.

need only be extracted and compiled. Compilation, analysis, and reporting can be done as absolute numbers and as indices. Cost figures should also be shown for product lines, independently managed organizational units and plants, and special projects and time periods.

Costs per unit normally show decreasing defect/failure cost with increasing efforts and costs for prevention and appraisal. With increasing effectiveness of the quality assurance program, total quality costs should decrease, thus contributing to profit.

Quality cost reports permit regular performance controls. Once irregularities are observed, causes can be more rationally analyzed and remedied.

Table 1-5. Determining a Suitable System Standard

1. If a major customer requires compliance with a published quality assurance system standard, obtain that document. Note that our Guide does not differ basically from any such standard, but adjustments will have to be made in order to comply with the standard.
2. If major competitors have a superior quality assurance system, assess and possibly adopt their standard.
3. If you want to establish your own system:
    a. Obtain the Generic Guidelines for Quality Systems (ANSI/ASQC Q94) from the American Society for Quality Control.
    b. Determine the problem areas and system elements you want to prepare quality assurance procedures for, considering factors such as major potential defect, prevention, and appraisal costs; competitor quality assurance; product and process maturity.
    c. Determine if our Guide is a sufficient information source. If yes, do not acquire any other standard for the time being. If no, proceed to next step.
    d. Select a standard from Table 1-4, preferably one of the ISO 9000 series.
    e. Assess that standard as an additional information source.
    f. Keep that standard for reference when designing the quality assurance system and writing individual quality assurance procedures.

2
Designing The Quality Assurance System

A company's goal should be to establish a suitable and effective quality assurance program. Sometimes this means reviewing and improving an existing program, other times, one has to start from scratch. We assume the latter situation. Moreover, the design task has to be done both in large and complex organizations and in small businesses. We address design in larger companies, but wherever possible, simplify the procedure for small businesses. The reader must augment the procedures in our Guide, or use the procedures selectively and adapt them to his or her own circumstances.

We have chosen to translate the overall issue and problem of designing into individual projects. All projects will be consistently directed toward the overall goal, embody the same principles, and remain interrelated and interdependent. In other words, the overall project divides into subprojects. Working out an action plan is the first project.

2.1 An Action Plan

QUALITY ASSURANCE PROGRAM ACTION PLAN

PURPOSE

The action plan describes and initiates approaches and projects for the establishment of a quality assurance program.

POLICY

Designing the quality assurance system is a project under the direct supervision of senior management. The quality assurance group will initiate and coordinate all the activities. All other groups will participate as requested and will be held responsible for the quality assurance and related actions in their areas of responsibility.

DEFINITIONS

A *quality assurance program* includes all the decisions and actions required to attain and maintain the quality of performance and output that satisfy customer and workmanship standards. *Quality procedures* are designed to guide and support this quality performance; they embody proper practices. Each project has its individual action plan, which is an integrated part of the overall action plan.

REFERENCES

Available information resources and technical aids should be applied in the action plan, and appropriate references to these resources should be made.

SCOPE AND APPLICATION

Both the overall and the individual action plans must have clearly defined goals and should be cost-effective. The action plan's scope, that is, the number of procedures, the areas and functions to be covered, the personnel to be involved, and so on, should be planned carefully and justified. The scopes of the initial and the designing stages of the action plan can be narrow, and gradually widened as progress is made. In small enterprises narrow scope and careful planning can overcome the lack of expert assistance.

Action plans should be prepared by well-trained and -qualified personnel and should be addressed to formally designated staff. Each plan must respond to a well-researched and -justified need before actions are assigned and the plan is carried out.

STEPS/METHOD

There are various approaches to designing a quality assurance system. The following outlines are examples of a comprehensive and a simplified action plan. The steps describe a general set of activities that follow each other sequentially. In practice, they must be further expanded in terms of individual activities and assignments. The steps can then be arranged and handled as networks. Each network has a start time and a deadline—the start and end node. The project plan and respective network, or schedule, stipulate details such as resource requirements, performance standards, milestones, work packages, and organization.

PLAN A (COMPREHENSIVE)

1. Appoint a program coordinator and establish a steering committee.

2. Prepare and conduct a management forum to inform all affected personnel about the quality assurance system. (Use outlines in Chap. 1.)

3. Prepare and conduct a companywide survey of existing quality assurance practices and procedures. (Use checklists; examples are shown in Tables 2-1 and 2-2.)

4. Compile applicable program standards and guides. (See Chap. 1.)

5. Determine the need for organizational, technical, economical, and procedural improvement of quality assurance (possibly using applicable standards and audit format).

6. Prepare and conduct a management forum to establish goals, policies, and strategies (document the results in a Quality Manual).

7. Establish or review the organization for administering the program (describe and document this in the Quality Manual).

Table 2-1. Checklist for Simple Quality Assurance Survey

Quality Assurance Activity	Current	Desired	Comments

What is inspected or tested?
    materials
    supplies
    work in progress
    end item
    processes/equipment
    measuring and test equipment
    drawings, etc.
    customer complaints, returns
    other			

Where is inspection/test performed?
    supplier
    receiving
    work station
    laboratory
    before delivery
    after sales
    after service			

What types of inspection or tests are performed?
    visual
    measurement
    destructive
    functional
    other			

What is the sample size that is inspected/tested?
    selected item
    prototype
    sample
    statistical sampling
    100%
    other			

Who inspects or tests?
    operator
    supervisor
    inspector
    manager/owner
    supplier
    customer			

Table 2-1. (*Concluded*)

Who decides on nonconformance handling?
  operator
  supervisor
  inspector
  manager/owner
  committee
  customer			

What use is made of the inspection or test results?
  planning of quality
  modification of quality
  process capability studies
  quality report/records
  performance evaluation
  other			

What are the general quality assurance actions?
  quality manual
  quality circles
  training workshops
  audits/reviews
  quality improvement projects
  motivational programs
  quality cost survey
  other quality surveys
  other			

8. Initiate, support, and coordinate procedure writing or review projects:
   a. determine principles and layout for procedures
   b. select and use procedures in this Guide as first draft
   c. conduct a pilot project
   d. conduct a workshop on procedure writing
   e. determine, assign, and support procedure writing project (use project management techniques, i.e., computer-based network analysis)
   f. review drafted procedures and incorporate those approved into the Quality Manual
   g. audit implementation of procedures
   h. recognize successful procedure writing project teams

Table 2-2. Checklist for Initial Survey of Quality Assurance Practices

1. Does the company have a quality assurance program?
2. Is it documented in a quality manual?
3. What standard does it or should it comply with?
4. Are quality/inspection plans prepared?
5. Are individual inspection procedures clear, current, and complete?
6. Is the inspection equipment in working order?
7. Are all inspection/tests identified?
8. Are inspection records reliable and useful for audits?
9. Is one person assigned quality assurance responsibilities?
10. Is the inspection function sufficiently independent?
11. Are the responsibilities for quality assurance clear and comprehensive?
12 Are inspection performances reviewed through audits?
13. Are audits properly planned and executed?
14. Are corrective actions taken properly?
15. Does the company control the quality of its supplies?
16. Are supplier audits conducted?
17. Is the inspection status on products clearly indicated?
18. Are processes controlled with regard to quality specifications?
19. Can any article by-pass required inspection?
20. Are statistical control methods applied?
21. In cases of nonconformance are proper actions taken?
22. Are nonconforming articles held in a restricted area?
23. Does the final inspection review all previous inspections?
24. Were all inspections carried out in accordance with the applicable standard?
25. Are the records complete and the inspection traceable?
26. Are handling, packing, and shipping controlled?
27. Are customer complaints recorded and properly handled?
28. Are quality costs compiled and analyzed?
29. Is senior management visibly committed to quality assurance?
30. Do programs for instructing and motivating staff exist?

9. Establish (review) the quality management information system (include quality cost accounting and reporting).

10. Establish infrastructure (budget, resources, procedure) for special quality improvement projects.

11. Audit the quality assurance system and submit a report to senior management.

12. Conduct a meeting for launching and publicizing the quality assurance program.

13. Request external audits and registration.

PLAN B (SIMPLIFIED)

1. Appoint a quality assurance coordinator.
2. Select a program standard or guidelines (see Chap. 1).
3. Compare requirements with current practices and take corrective action, if necessary.
4. Formulate practices as procedures. (Use prototype procedures in this Guide as a first draft.)
5. Complete quality manual (policy statement, records, forms, coordinator/organization).
6. Review/audit implementation.
7. Publicize quality assurance program (possibly request external audit and registration).

TECHNICAL AIDS

In a quality assurance program, project planning and control techniques and related computer software can be used. These programs for mainframe or personal computers require various data inputs, such as name of activity/subproject, immediate predecessor, estimated duration, and other attributes. Various output reports can then be generated showing "critical path(s)," "slacks," etc. These reports should meet the actual information requirement.

2.2 Project Planning and Control

Establishing a quality assurance program is a complex project, where designing the program is the first integrated subproject. In order to manage all these sequential and interdependent projects in a consistent and effective manner, a general project management guideline becomes necessary.

PROJECT MANAGEMENT GUIDELINE

PURPOSE

This guideline will assist you to plan and control consistently and effectively all individual projects undertaken when establishing a quality assurance system.

POLICY

Designing, implementing, and maintaining the quality assurance program are performed as integrated projects. Suitable and cost-effective project management methods should be applied, and computer-based planning and control systems should be used.

APPLICATION

The general project management guideline is addressed to project managers and teams. Those initiating and supervising projects might augment or modify this guideline.

REFERENCES

Guidelines for quality improvement projects and procedure writing; manuals for project management computer software to be applied; and any other available project management guidelines can be used for reference.

DEFINITIONS AND PRINCIPLES

A *project* is a complex task that has a definable beginning and end and involves many interrelated activities. The *project's goal* is defined in terms of expected outcome, time, and investment (costs). The project's scope, uniqueness, and relative degree of difficulty, and the qualification of project manager and team all influence the approach one should take to accomplish the project's goal. Relatively simple tasks and assignments should not be handled as projects.

The major principles for project planning and control are as follows:

* Clear definition of
    — the project's objectives and the outcome,
    — the project's major milestones,
    — all the project's tasks and assignments and their interrelationships,
    — relationships to other related projects.
* Organization of project team.

- Budgets and resource availability.
- Schedules and computerized project reporting and control schemes.
- Interface of project management and departmental functions.
- Frequent meetings and communication.
- Recognition of contributions to the project's success.
- Timely progress review and updating of project plan.

The following questions should be answered by the project manager:

- Is the project meeting the planned schedule and cost estimates? If not, what are the deviations and their causes?
- Is the outlook for meeting the schedule and their cost estimates improving or getting worse, and why?
- Is the team adequately qualified and cooperating? If not, why not?
- What problems are being encountered? What corrective actions are necessary and have been taken?
- Is information for decision and action sufficient? If not, can it be made available?

STEPS/METHOD

A project moves through the definition, planning, scheduling, and control phases. The individual steps are as follows:

1. Definition phase
   a. Define and clarify the project's objectives and its major features.
   b. Break down the project into subprojects or work packages.
   c. Determine the subprojects' interrelationships.
   d. List resource requirements and organizational responsibilities.
2. Planning phase
   a. Determine individual activities for each subproject.

b. Determine each activity's data input to the computer-based information system, such as time estimate, immediate predecessor(s), resource requirement, and performance standard.
c. Assign tasks and clarify responsibilities.

3. Scheduling phase
   a. Prepare the plans for the project and subprojects in a computer-generated-report format.
   b. Assess the tentative schedule and critical sequence of activities (critical path) and modify them as necessary.
   c. Communicate the plan and schedule and get final approval.

4. Control phase
   a. Initiate the plan and set the schedule.
   b. Review the project's progress and take corrective action when necessary.
   c. Report on outcome.

TECHNICAL AIDS

This Guideline assumes the use of computer software for project planning and control; there are packages for both personal computers and mainframes. Some examples are Harvard Project Manager, MacProject Scheduler, Microsoft Project, Superproject, Timeline, Pertmaster, and LOTUS 1-2-3. Some of the advantages that computer software use offers are as follows:

* Convenient integrated planning, implementation, and progress control.

* Operations manuals and error diagnostics.

* User-friendly data input and processing.

* Generate tailor-made reports.

* Handle several projects and the interdependencies concurrently.

* Allow standardization and simplification.

* Provide documentation for performance evaluation.

- Facilitate continuity in projects, training, and orientation.
- Improve personnel cooperation and coordination through the availability of more factual and up-to-date information.
- Identify troublespots and provide flexible responses.
- Relieve project manager and team of administrative duties.

The following scheme provides a detailed overview of project phases and steps:

Phase		Step
1. Preparation	11	Determination of goal/task
	12	Team/organization
2. Improvement alternatives	21	Development
	22	Appraisal
	23	Testing
3. Decision	31	Evaluation of alternatives
	32	Selection/deciding
	33	Planning/implementation
4. Implementation	41	Preparation of schedule
	42	Execution
5. Control	51	Performance
	52	Reporting
	55	Corrective action

Substeps and their related questions are:

Substeps		Questions
111	Starting point	where to start
112	Status quo	what is the current situation
113	Goal, purpose	what should be
114	Cause	why
115	Constraints, budget	what are the limits
116	Assignment	who
121	Format	how
122	Responsible department	which
123	Desired deadlines	when
124	Participation	who
211	Conventional alternative	how was it done before
212	Similar tasks	how is it done in similar areas
213	Possible combination	what are partial solutions
214	New alternatives	what are innovative ways
215	Procedural requirement	what measures must be taken
216	Sequences, priority	in what sequence and ranking

Substeps		Questions
217	Technical aids	what aids are necessary
218	Process planning	what is the process/approach
221	Dependencies	what are the dependencies
222	Determinants	what are the decisive factors
223	Linkages/relationship	what are adjacent areas
224	Marginal cases/norms	what are limits
231	Experimentation	what are practical results
311	Weaknesses, defects	what are shortcomings
312	Business policy	what are relations
313	Optimal alternative	what appears optimum
321	Selection	who is to decide
331	Preparing execution	how to insert the project
332	Work packages	what are the activities
333	Work assignment	who is to be in charge
334	Relationships	in what sequence
335	Location	where
336	Procedures determination	with whom to adjust/coordinate
337	Final procedure	how
338	Communication	with what resources, aids, techniques
339	Forms	with what organizational aids
411	Schedule/network	when/what to start and complete
421	Starting point	who is responsible
422	Progress control	goals/milestones achieved; initial
511	Performance/quantity	what progress achieved
512	Performance/quality	what progress achieved
513	Performance/time	was deadline observed
514	Costs	what were the relative costs
515	Effects, negative	what were the negative effects
521	Reporting	what, to whom, in what form
531	Task adjustment	what must be modified
532	Additions, deletions	what must be added/deleted in the remaining project part

2.3 Quality Improvement Projects

When designing a quality assurance program, one essentially deals with quality improvement—existing quality assurance practices need to be better formalized, systematized, and structured. Properly managed projects for rectifying obvious weaknesses in current quality assurance procedures usually have a good chance for success.

We shall follow-up this section on quality improvement projects with those for procedure writing. Quality improvement projects normally lead to the review or establishment of proper procedures.

GUIDELINE FOR QUALITY IMPROVEMENT PROJECTS

PURPOSE

This guideline is to help with the effective and efficient implementation of quality improvements through proper project planning and control.

POLICY

Quality improvement is a continuing task that should challenge and involve the entire company. In order to provide an adequate infrastructure for quality improvement, a project management approach should be applied; gradual improvements can also lead to an improved quality assurance system.

APPLICATION

This guideline is to assist managers, supervisors, team leaders, and those recommending quality improvements to become involved in their respective projects.

DEFINITIONS AND PRINCIPLES

Quality improvement is perceived from the customer's perspective and results from joint decisions and actions of management and staff, and their coordinated effort and cooperation. Improvements result from creative and proficient project planning and control.

Some quality improvement principles follow:

* Gains through quality improvement projects must be measurable and visible.

* Projects must meet all the principles of sound project management (planning, implementation, and performance control).

* Projects must be arranged by priority, which is determined by need and the existing capacity to optimally fill that need.

* Initial projects should identify subsequent promising projects and assign them priorities.

* The listing of projects and their study/survey must include all functional areas and departments in an organization.

* The total quality improvement of all projects should be higher than the sum of the individual gains.

REFERENCES

References to study are the Project Management Guideline and the Action Plan Procedure outlined in the preceding sections.

STEPS/METHOD

1. Identify the areas where quality improvement is needed.

2. Survey operations and solicit recommendations for quality improvement.

3. Prepare a list of quality improvement projects. Each project defines a specific objective, with deadline, team members, resources required, and major milestones.

4. Rank individual improvement projects by their urgency and the potential improvement's importance and benefits.

5. Establish the first project team and prepare the project plan. The main objective of this project is to remedy major causes of poor product/service quality and to initiate companywide quality assurance.

6. Implement the first quality improvement project with sufficient publicity. Audit the project's progress, completion, and achievement. Provide for appropriate recognition of outstanding contributors.

7. Prepare an overall development plan for the companywide quality system. Appoint a project team. Use the applicable quality-related standard as the major document for planning the system. This standard outlines the functions the system

will have to incorporate; each function of the system can itself be a project.

8. Carry out the projects either sequentially or concurrently, with audits performed at important milestones and upon project completion.

9. Establish a Quality Management Information System that includes and coordinates procedures, manuals, reporting, planning, auditing. (See our Prototype Procedure, Chap. 3.) The system should be computerized as much as possible. Quality cost accounting must become an essential part of this information system.

10. Establish a system of regular meetings devoted to quality improvement and problem solving; Quality Circles could be adopted. These meetings should be incorporated in the Quality System; for instance, reports could be communicated to personnel responsible for planning and auditing.

11. Design an integrated planning and auditing system for the continuous supervision and improvement of the Quality System. The audit system facilitates independent control of operating units and should extend to both suppliers and customers.

12. Once the Quality System is well established and is performing satisfactorily, it should be integrated with other functional systems, and possibly registered. Integration and registration provide essential visibility and acknowledgment for the Quality System.

This approach can of course be simplified, in particular, in small businesses. However, individual quality improvement tasks and projects eventually should be integrated into the main quality assurance program.

TECHNICAL AIDS

Various software packages and associated manuals as listed in Section 2.2 should be applied. Certain quality improvement projects can be organized as Quality Circles; see publications in the catalog of the American Society for Quality Control. Prototype procedures in this Guide on statistical methods, for instance, can assist in technical matters.

2.4 Procedure Writing Projects

Designing a quality assurance program consists essentially of drafting procedures. These guides and the compliance with them lead to quality improvement. This desired outcome, however, requires that the procedures are well planned and carefully introduced into the company. The following guideline is to assist in writing sound quality assurance procedures.

GUIDELINE FOR PROCEDURE WRITING

PURPOSE

The guideline for procedure writing will help team members attain uniform and suitable quality assurance procedures that effectively implement applicable policy/strategy and system standards.

POLICY

Quality assurance procedures govern quality assurance activities whenever the need for and applications of these procedures are properly studied and clarified. With program standards one can determine what procedures are required; however, additional procedures can be instituted when needed.

Supervisory management will support and approve procedure writing projects and the resulting procedures. All personnel concerned with a specific procedure should participate in its drafting. Procedures are reviewed and maintained through audits.

APPLICATION

This Procedure Writing Guideline is to be adopted by all those project teams and individuals formally assigned to prepare a quality assurance procedure.

REFERENCES

Standards for Quality Management Systems as listed in Chap. 1. Standard operating procedures in Frank Caplan, *The Quality System,* available through ASQC. Prototype procedures of this Guide can be used as a first draft.

DEFINITIONS AND PRINCIPLES

A *procedure* is a formal and mandatory directive and guideline for work performance that provides all the necessary information and performance criteria. Supervisory management plans, approves, and audits these procedures. The components of a procedure are purpose, policy, application, references, definitions, performance steps/sequence, and criteria. A brief explanatory note can be added. A procedure differs from practice, policy, instruction, etc.

The general principles for procedures are:

* The writing is arranged as a project with a specialist as coordinator.
* Those addressed by the procedure participate in planning, drafting, and implementing it.
* Training is provided.
* Procedures are reviewed regularly or upon request.
* The procedures are prepared in conjunction with standards and guidelines.
* The final procedure must be:
  — operational and acceptable,
  — consistent with policy and needs,
  — in compliance with standardized format, see Technical Aid Subsection.
  — optimizing operator's independent planning and control (self-inspection).

Organizational and technical specifics are:

1. Procedures will be initiated by senior management; they can be requested by supervisors and quality assurance personnel. A rough draft should be submitted. Auditors should

evaluate procedures and compliance and should report to senior management.

2. Establishing the procedure will be assigned to a competent body that must follow guidelines and other directives from supervisory management.

3. The procedure will be named, coded, and referenced. A draft will be prepared. This draft is to be assessed by all concerned and to be approved by supervisory management through signature and date.

4. The procedure's table of contents has the following headings in sequence: Title, Purpose, Policy, Application, References, Definitions, Steps/Methods/Criteria, and Attachments (forms). Deviations have to be approved by supervisors and management.

5. The procedure's status must be clearly signified on the form. A draft might have to indicate the various writing stages. A procedure can be declared: "tentative," "under review," "new," or "optional." Documents on invalid procedures must be withdrawn and destroyed.

6. The procedure must be signed by supervisory management, and changes must be dated.

7. The procedure is to be documented in a manual along with other related procedures and is to be distributed and communicated effectively.

8. Procedures must be readily available to all persons affected by it.

9. All procedures are to be audited annually or earlier when requested.

10. Procedure writing projects must be given a deadline and adequate time and resources to be accomplished.

STEPS/METHOD

1. The proposal is submitted to the supervisor, manager, and quality assurance specialist (coordinator).

2. The need for and the application of the procedure are researched and the result is approved (or disapproved and the project is abandoned).

3. The draft procedure is prepared using standards and prototypes.

4. The draft procedure is communicated, discussed, and modified, if necessary.

5. Training is provided.

6. A tentative procedure is introduced for testing.

7. The procedure's implementation is audited and its results are reported.

8. After final approval, the procedure is officially introduced.

9. The procedure and the compliance with it are audited according to plan or upon request.

TECHNICAL AIDS

In our Guide, guidelines for project planning and control are designed to manage procedure writing. The prototype procedures outlined in the next chapter can serve as a first draft.

Applicable standards will outline quality assurance definitions and principles:

* "Operational procedures coordinating different activities with respect to an effective quality system should be developed, issued and maintained to implement corporate quality policies and objectives. These procedures should lay down the objectives and performance of the various activities having an impact on quality, e.g., design, development, procurement, production, and sales." (ANSI/ASQC Q94-1987, ISO 9004)

* "Once the required type of quality program has been selected, the program should be compared with a list of actions, the purpose being to identify additional activities which are necessary. The next step is to draft procedure outlines for all activities, existing and newly required, with the assistance of

those who are and will be involved. Agreement, approval, and support at management level is essential." (CSA Z299.0)

PROCEDURE FORMAT

Title: Brief, descriptive, possibly repeat title found in the Standard; attach code.

Purpose: Defines rationale for and objective of procedure; be specific.

Policy: Defines general rules and principles; translates strategy.

Application: Defines function of and persons addressed by the procedure; time and conditions for its application, rules governing deviations and approvals, and assumptions made.

Definitions: Terms that must be known by the performer relative to the procedure; references to glossary might be made.

References: Standards, other directives, specific work/test instructions, technical specifications, forms.

Steps/Method: Outline of performance sequence without prescribing details of each step in technical terms. The prototype procedures in this Guide are not detailed work instructions; use action words.

Forms/Technical Aids: Forms for records and reports must be standardized in heading, body, identification/reference, cross references, and use of codes. Technical aids are standards, guidelines, prototype procedures in this Guide, and illustrations.

Explanatory Note: Brief outline and augmentation on subject matter of respective procedure addressed to managers who are nonspecialists in quality assurance management.

3
Quality Assurance Procedures: Subsystems

3.1 Design Assurance and Contract Review

Design assurance assumes that a new product or service has to be developed and that detailed quality specifications which facilitate quality control have to be established. The design review evaluates and audits all design activities for compliance with established procedures and for attaining milestones and design objectives. A market readiness test is often part of the design review and determines the acceptance of the product or service in the market place.

Contract review refers to a subcontract negotiated with a customer. It is to confirm the ability to meet contract obligations; it is usually undertaken before the contract is signed.

DESIGN ASSURANCE

PURPOSE

The product or service or both must be designed to meet the customers' needs, the conditions of applications, the appropriate levels of price and safety, and the adherence to applicable standards.

POLICY

Marketing and Design Engineering in cooperation with Quality Assurance must ensure that quality is assessed in all phases of design.

APPLICATION

This procedure and all referenced, detailed work instructions are to be followed in all specified design inspection points by staff members responsible for product and service development and design assurance and qualification.

DEFINITIONS

Design is the result of researching and surveying the customers' stipulations and specifications for products and services, determining the product's or service's dimensions or attributes, developing and field testing prototypes, determining technical and resource requirements, and documenting and finalizing the product or service.

Design assurance embraces all decisions and activities directed toward establishing and meeting the quality objectives to the customers' satisfaction. A thoroughly planned and documented quality specification facilitates the design's quality control. Approved attributes and dimensions of a service or product must be documented.

REFERENCES

References include customer supplied drawing and specifications, mandatory product and service-related codes and standards, market research reports, failure effect and mode analysis reports, and product briefs. Quality Management System Standard, i.e., ANSI/ASQC Q94-1987, American Society for Quality Control, Milwaukee, WI.

STEPS/METHOD

1. Assign the responsibility for design assurance for each product or service, contract, order, etc., subject to new design or design review.

2. If applicable, assess either each order received from Marketing/Sales or new products or services to be offered with regard to applicable quality standards and productivity. Specifications must be clear and complete. Design and Production Engineering must approve the design specifications.

3. For major orders specified by customers rather than through Marketing/Sales evaluate contract specifications; see "Procedure for Contract Review" in the final subsection. In addition to verification as mentioned in item 2, the specifications should include a description of the application and performance conditions for the product or service and a detailed Quality Plan (see the "Procedure for Contract Review"), which should be approved by Design and Production Engineering.

4. Establish design project schedule with intermediate milestones, and delineate the schedule's major activities and goals.

5. Verify that the design process has been adequate and in conformance with directives and standard design procedures. Documentation at major inspection points must be complete and clear.

6. Verify that customer and regulatory bodies have approved the design.

7. Drawings and other documents describing the design should be inspected for completeness and compliance with directives and standards.

8. Verify that design documents have been communicated and approved by Production and other departments responsible for conforming to the design.

9. Verify basic supplier and production readiness.

10. When applicable, verify that customers and regulatory bodies have formally approved final design documents.

11. Conduct independent testing of the product or service for design qualification and validation.

12. If applicable, conduct an independent and comprehensive Design Review/Audit verifying that the customer's needs and specifications; the product application performance; the process performance; the supplier conformance; the product or service marketing and distribution; and the product

reliability, availability, and maintenance have all been met. (See "Procedure for Design Review.")

13. Check market readiness test, if applicable, and all customer-directed information material (software) associated with the product or service. Check also the packaging, shipping, distribution services; installation; and other customer services

14. Approve design release and market readiness.

15. Review or audit product performance, customer feedback, and design requalification.

16. Verify the proper planning and control of any design requalification or any changes in design and the respective documentation.

17. Establish, or verify, design assurance planning and control, including a review or audit of compliance with the preceding design assurance procedure. Complement this Procedure with detailed work instructions and respective cross references, as applicable.

FORMS

The forms required for this subsection are work order, engineering change notice, design factor listing, and product briefs.

EXPLANATORY NOTE

The marketing department is mainly responsible for determining the product or service's quality characteristics and the customer's requirements; marketing criteria affecting the customer's satisfaction; and communication among customers, distributors, regulatory bodies, etc. When individual customers contract for major items and projects, the marketing department establishes and maintains contact in matters concerning quality. The marketing or design engineering departments or both should inform other departments about customer requirements, and should inform the customer about the capabilities of their organization.

Specifications should include packaging, references to the major defects and failure risks that might be encountered, customer instructions and services offered, testing methods, and standards.

The frequency and thoroughness of the design review depend on the risks likely to be encountered and the impact of any defects or failures. The results of reviews and the methods used to perform them should be recorded. Design reviews can include testing by and with the users and customers, comparisons with alternative designs, inspection and testing feasibility, supply quality assurance, customer service, process capability, workforce capability, field testing, and information feedback and corrective action.

The design review should be performed by independent, competent personnel, who are familiar with the technical and economic aspects of the design and the future marketing strategy. Standards for quality auditing also apply to design reviews.

All documentation of design planning and review should be coded and filed.

Reviews for design requalification should be conducted periodically or when such indications for a review as design changes, customer complaints, major defects or failures, or new technology make redesigns necessary.

DESIGN ASSURANCE, CHANGE CONTROL

PURPOSE

Changes to drawings, specifications, and other related decisions and their documentation must be planned and approved; obsolete documentation must be eliminated.

POLICY

Staff members and other persons concerned with the design and its implementation must always have the latest design documentation and information. Anyone requesting changes must initiate these changes in accordance with the proper procedure.

APPLICATION

This procedure applies to all persons having authorized access to design documentation. Other persons will request or suggest changes through supervisors or others authorized to receive such communications.

DEFINITIONS

A *drawing* is any illustrative or schematic description of an item, or part thereof, used to communicate a design and to facilitate the design's production. A drawing can be associated with explanatory documents; it can also be a product of computer-assisted design (CAD).

A *specification* is any quality characteristic—attribute or measurement—that is officially stipulated and approved. The totality of product- and service-related specifications represents quality that is designed to meet the customers' needs and requirements under defined conditions for application and performance of the item in a safe, reliable, and economic manner.

Change is any alteration of a specification in drawings or other design documents that has been approved by an authorized person.

REFERENCES

References for this subsystem include system standards; applicable technical standards; design assurance; change and design review procedures.

STEPS/METHOD

1. Designate a department or person authorized to receive, process, and decide on all requests for change.

2. Requests for change may be made by anyone, but should be made by staff members when improvements are necessary. These requests can take the form of a simple "Trouble Report." A request should be documented in a "Request for Change" form.

3. All requests for change including any comments by the receiver of the request must be sent to the authorized person or department for a decision and action.

4. A request for change is assessed on its effect on quality and is processed by Design Engineering, Quality Assurance, Marketing, and the other departments concerned. In emergencies caused by defects or comprehensive product recalls, special procedures and directives must be available. For major recommended changes a design review should be conducted. Customers and regulatory bodies might also become involved.

5. Changes, when approved, must be made in all design documentation, and should be disseminated quickly. Prior notice of an impending change might be given.

6. After a major or several changes, drawings and associated design documentation must be reissued and the obsolete ones collected and destroyed.

FORMS

Forms include Request for Change Note; Change Note; "Trouble Report," which includes space for remedial/corrective action or design change.

EXPLANATORY NOTE

Whenever a formal design document, such as drawings or quality specification listings, is prepared and becomes a contractual obligation, changes must be properly controlled. Anything that is not documented is left to the discretion of the production or operation department; workmanship standards, skill, and tradition might govern the quality of a product or service. In such situations changes will take the form of adaptation to communicated, perceived, or anticipated customer requirements and expectations.

Changes in drawings and specifications normally require concomitant alterations in work instructions.

DESIGN ASSURANCE, DESIGN REVIEW

PURPOSE

As in audits of operations or systems, designs must be independently evaluated as to how they comply with their specifications and procedures. Moreover, weaknesses or improvements are to be detected or implemented by this design review at the crucial development stages and at the end stage.

POLICY

New designs and those being used must be reviewed periodically or when indications or incidences require a review. The review is to verify compliance and validate the design.

APPLICATION

This procedure is applicable to persons and functions or teams formally assigned to undertake a design review; to persons involved with design development and change activities; to the quality assurance department; and to persons initiating design development and change.

DEFINITIONS

Design review is a formal, independent evaluation of a product's or service's design, including its development, implementation, and requalification stages, to verify compliance with specification, standards, procedures, and directives, and to improve design effectiveness. It is neither a regular inspection of design activity nor a project progress meeting. The objects of the design review are all items and aspects looked at from the customer's point of view; major product specifications for safety, environmental impact, etc.; and major production/process specifications.

Design effectiveness is the degree of customer satisfaction a design creates within the limitation of various price, technological, social, and economic factors. Design verification measures the design effectiveness.

Design baseline is the result of the final design review and represents the approved design ready for market testing and initial production.

REFERENCES

References for this subsystem include applicable and known standards governing the design; Quality Assurance Design Review (DOD 64245.7); and "Best Practices Guide," U.S. Navy document NAVSO-P6071. Design assurance and change procedures.

STEPS/METHOD

1. The person or department performing the design review will be appointed by senior management; this appointment is made from a pool of competent and independent persons. A review team with one person as leader should be appointed for comprehensive and complex design and review projects.

2. The objective and scope of the review must be established, documented, and communicated. Reviews might be integral parts of the design procedure or invoked by special needs or requested.

3. A review plan showing individual review activities, tests, verification, place, time, function, team members, requirements for cooperating persons or departments must be prepared for complex reviews.

4. A review plan should be communicated to and approved by supervisory management to ensure that all aspects of the design and design activities are in accordance with the stated objectives of the review.

5. Reviewers should receive training in, be briefed about, and have access to design directives, documentation, etc., in preparation for the review. Working papers that facilitate a reliable and effective review should be prepared.

6. Reviewers announce the start of and then conduct the review in accordance with the established plan and other

review guides. The tests conducted depend on the review's objective, the current development stage, the relative significance of specification and verification, and the applicable standards. The review should include the major factors related to customer satisfaction and application requirements, and production or process specification and capabilities. The referenced and applicable standards further explain details of the design that are subject to review.

7. The reviewers draft a report and communicate and discuss the draft with Design Engineering, Quality Assurance, and other persons or departments involved in the design activity.

8. A written report and the main test documentation are submitted to senior and supervisory management personnel. Other people, such as customers or regulators, might receive a copy of this material through senior management.

9. Senior management decides on any follow-up of the design review, files the report, and after completion of the review, disbands the review team or relieves the individual reviewer.

10. An official production release is prepared and disseminated.

FORMS

The forms needed for a design assurance review are design review request, review plan, working paper, and, for a major review, a form for the review report and follow-up.

EXPLANATORY NOTE

A design review can range from a very simple verification to a complex series of interdependent reviews. The governing principle is that the integrity of the design is ensured to the extent of the customer's demand and willingness to pay for it and the mandatory standards and directives that stipulate it. The review is neither a general, informal survey of design development that is part of normal managerial supervision nor a detailed testing and inspection that is an essential element of normal or prescribed design development and verification procedures.

Basically, design reviews follow principles and procedures for quality system audits; in fact, they can be organized as an integral part of these audits. Design reviews, however, focus on a product or service that is in the design or design-modification stage. There are many risks and uncertainties in design development—as is true for any other research project, innovation, or technical development—that do not always allow standardization of design features. Such a lack of standards suggests a review rather than a stricter compliance audit; only design procedure compliance can be audited. A market readiness test applies when an individual customer is not known at the design stage.

REVIEW OF A CUSTOMER'S CONTRACT

PURPOSE

Before agreeing to a contract and accepting its obligations, particularly concerning quality specifications and quality assurance, a special contract review must ascertain that there is adequate understanding of its specifications, documentation, quality assurance stipulations, and requirements, and the organization's capability for compliance.

POLICY

No contract should be signed and finalized that has not been checked concerning the quality and quality assurance requirements and the organization's ability to satisfy the customer.

APPLICATION

This review applies to all staff members explicitly authorized to negotiate and sign contracts with a customer.

DEFINITIONS

A *contract* is a legally binding agreement between two parties to accept and meet specified obligations. In the context of this procedure and quality assurance, a contract review focuses on quality and quality assurance obligations.

REFERENCES

The applicable references are design assurance and change procedures; statistical process control procedures; design review procedure when part of a contract review; and applicable standards.

STEPS/METHOD

1. A person or department is appointed for contract review: either a person authorized to negotiate and accept a contract or a specially assigned reviewer.

2. Determine what customer specifications are stated in the tender documents or other communication.

3. Determine the general and specific offers, claims, specifications made in advertisements, solicitations, etc.

4. Verify the completeness of the quality and quality assurance requirement, and whether these specifications are understood by both customer and supplier. Specifications and all obligations must be clear and adequately defined and documented.

5. Verify that all major obligations are in compliance with applicable codes and standards; if necessary, obtain official confirmation.

6. When a disagreement or a different interpretation of obligations occurs, obtain a clarification, which might require negotiation or special investigation.

7. Verify that production capabilities exist to meet the quality and quality assurance requirements. This verification might require testing, modification of the quality specification, or negotiation to modify the quality specification.

8. Communicate the results of the contract review to all affected parties.

FORMS

The form required for a contract review is a contract review checklist; a statement in regular sales contract concerning quality assurance might suffice.

EXPLANATORY NOTE

A contract review precedes the formal acceptance of contractual obligations. Legal counsel and quality assurance experts should be consulted in major contract negotiations.

When salespeople are normally authorized to close contracts for individual items, supervisors must validate such contracts and verify the acceptability of the quality and quality assurance obligations. Particularly with new designs and custom products or services, a special and formal contract review is a prudent practice. Product liability and customer dissatisfaction can often be traced to misrepresentation and misunderstanding during contract negotiation, where price rather than quality all too easily dominates.

3.2 Quality Plan or Inspection Plan

QUALITY PLAN OR INSPECTION PLAN

PURPOSE

The quality or inspection plan ensures that the product or service design's specification is maintained by the planning and control of explicit quality-oriented activities and that documentation concerning that particular product or service is complete.

POLICY

For any product or service a quality plan is to be prepared that initiates and guides quality assurance activities. Detailed inspection and test instructions complement the quality plan.

APPLICATION

The quality plan is applicable to personnel and departments responsible for the quality assurance and management system or program and those staff members assigned to the quality assurance, control, and inspection responsibilities.

DEFINITIONS

A *quality plan* is a document describing all product- and service-oriented activities including function, specification, stage in production, test or inspection method, and decision rules. A quality plan, when describing only inspection points and methods, is called an inspection plan or inspection checklist.

A *major defect* is one that renders a product defective and unusable.

REFERENCES

Applicable references are standards, customer requirements for quality plan, and regulatory body requirements.

STEPS/METHOD

1. Collect all drawings and design documents and check their completeness and accuracy.

2. Collect the production and work plans, including flow process charts, layout plans, process and workplace specifications, workmanship standards, schedules, and materials handling.

3. Assess the production and work plans from a quality and quality assurance perspective. These plans must be useful in meeting quality and quality assurance specifications.

4. Determine those specifications of the design that must be specially ensured for conformance and performance; classify them as primary (major) or secondary (minor) or according to some other classification (ABC).

5. Obtain approval from senior management and other affected personnel for the specification checklist that has been prepared.

6. Determine for each end item's specification the most appropriate inspection points. Prepare a production or process flow chart indicating these inspection points. Assess the inspection for economy; avoid unnecessary repeat inspections or other redundancies.

7. For each inspection point prepare an inspection or test procedure showing the test method and equipment to be used, inspection location, inspection personnel, decision criterion, nonconformance handling, and documentation to be produced. Special inspection or test procedures might be cross-referenced in the Quality/Inspection Plan. The inspection must be cost-efficient and reliable.

8. Determine special quality assurance points and methods for suppliers and subcontractors.

9. Determine supervision and control of testing or inspection; indicate points where such controls are to be performed and by whom. Allow for self- or roving control or both.

10. Coordinate quality or inspection plan with other production controls, such as inventory control, time schedules, and performance control.

11. Draft the Quality Plan and submit it for supervisory management approval. Customer and regulatory bodies might also have to approve the quality plan; check with the applicable standards.

12. Implement the Quality Plan; inform and train persons in charge of tests and inspections; allow time for learning the procedures.

13. Audit compliance with the Quality Plan and assess its quality assurance effectiveness.

14. Initiate any needed modifications to the Quality Plan and control these changes. Coordinate those changes with other design and production changes.

15. File the Quality Plan and standardize it for repetitive and large-scale production; maintain its cost-effectiveness.

FORMS

The forms needed for a quality or inspection plan are specification checklist, Quality Plan, and inspection or test plans for each inspection point.

EXPLANATORY NOTE

Prerequisites for preparing a Quality Plan are the completion and approval of the design and production plans. Modifications, particularly to production plans, could be required by quality assurance planning. Design completion includes a design review; the review report provides some guidance for the Quality Plan. Periodic quality audits will monitor the Quality Plan's effective implementation and any possible modifications to it.

The Quality Plan, as opposed to an Inspection Plan, covers production and performance stages in the product's or service's life cycle more comprehensively and is also conceptually more defect-preventive than -corrective. An inspection is a special and additional independent verification, sometimes assigned as self-control to operator or supervisor rather than to an inspector.

The Quality Plan, as opposed to a Quality System or Quality Management System, is product and service oriented. The Quality System is therefore more general and basic, since it embraces all products and services of the organization. A special quality assurance department might be responsible for the System/Program, while the Quality Plan is prepared by the design/production department in cooperation with Quality Assurance. The hierarchies of the Quality System and Quality Plan normally have individual inspection and test plans at the lowest level, which are usually assigned to the operational function.

The more responsibility for quality and quality assurance is delegated, the more supervisors and senior management must develop and rely on quality auditing for control and leadership.

The classification of specifications as major (critical) or minor (less critical) is done by taking into account the effects and impact of defects related to each specification. A "major defect" would critically affect the satisfactory and safe performance of a product being used by the customer for the intended purpose or the usefulness of a service rendered. Such division facilitates effective use of resources for quality assurance and cost-effectiveness.

The Quality Plan and its preparation and implementation can cover many different subjects and have many different forms. Some of the determinants of the subjects covered and forms used are the nature of the design and production plan; the requirements of customers; the applicable standards; the knowledge and experience of the staff; the required degree of quality and assurance; the available resources; the maturity of the design and production process; the technology available; and the strength of the competition. The Quality Plan is integrated with other product-, service-, and process-oriented plans.

3.3 Supply Assurance

The production and quality plans for a particular design indicate the external supplies and subcontracts required. In conjunction with a production capability study, a make-or-buy analysis might be performed. Supply assurance consists of supplier relations, purchase order and subcontract control, and receiving inspection.

SUPPLIER RELATIONS

PURPOSE

A list of suppliers and subcontractors must be prepared and maintained on the basis of general supply requirements and the supplier's availability, proven supply and quality assurance capabilities, and, if possible, actual performance record.

POLICY

Whenever possible, purchases and subcontracts should be made only with suppliers that are listed as "qualified." Any exceptions must be approved by the quality assurance department.

APPLICATION

This subsection applies to all persons or functions authorized to request, negotiate, and contract for supplies or subcontracts and to those responsible for production and quality plans.

DEFINITIONS

A *supplier* is a business or organization, operated and managed independently of the purchasing organization, that is prepared and able to accept and carry-out contractual obligations with conformance to specifications.

A *subcontractor* is a supplier of "nonshelf" items that must be produced in conformance with specifications for production processes and with the required quality assurance.

A *qualified supplier* is one who has requested and undergone an inspection or audit by the potential purchaser or an agent of the purchaser, has been approved, and has been listed by the purchaser accordingly.

A *supplier inspection or audit* is a formal and systematic evaluation of a supplier's ability to ensure quality to the satisfaction of a purchaser. The evaluation is commensurate with the relative need and importance of supply assurance, which is to be decided by the purchaser.

REFERENCES

Applicable references are standards for Quality Management Programs, Procedure for Contract Control, and the codes and directives of regulatory bodies.

STEPS/METHOD

1. Define the supplies that can be procured from unlisted suppliers; also list any exceptions to this list. A special procedure must assess the quality assurance of the supplier, define the purchaser's and supplier's respective obligations in the contract or order, and improve receiving and any source inspections. An unlisted supplier can be listed after a satisfactory performance.

2. For all major product and service designs and their respective production and quality plans determine the supply and subcontract needs. Categorize these, as with the specification and defect categorization, into major and minor supply assurance cases.

3. Verify that make-or-buy analyses have been performed and that the results for major supplies or subcontracts have been recorded.

4. Prepare documents for soliciting and receiving requests from suppliers who want to be evaluated as "qualified suppliers." This might be done for the major supplies or subcontracts only.

5. Assess current and past suppliers with satisfactory records for listing without an inspection or audit and inform them about the listing.

6. Prepare a program for reviewing or auditing the quality assurance capability of potential suppliers.

7. Conduct "desk reviews" for suppliers of major items and contracts first, ranked by priority. A "desk review" assesses the supplier's application, references, and general record, and, if available and required, the quality manual. Determine suppliers that must be inspected or audited.

8. Conduct in-plant audits or more narrowly defined product-related inspection.

9. Prepare a list of suppliers qualified to meet explicit and known quality and quality assurance specifications. Inform the suppliers. Those who are refused listing should be given a reason; others might be listed tentatively.

10. For each qualified supplier, or possibly only for major suppliers, prepare an individual record showing prelisting review data and supply performance.

11. Disseminate the "qualified supplier list" internally and get the list approved by senior management.

12. Assign maintenance of the list to a qualified person or de-

partment; conduct audits and requalify suppliers periodically. Make any additions to or deletions from the list in accordance with the preceding steps. Suppliers should be informed when they are deleted from the list.

FORMS

The forms required for this subsystem are as follows:

* *Supplier list*: name, etc.; items normally supplied, major–minor categorization; prelisting evaluation and results; supply contracts and performances; in general terms, major status changes.

* *Supplier record*: name (see supplier list entries); specific data on contracts and performance; agreements; observations; important correspondence or conversations; and references.

* *Desk review*: questionnaire for evidence of quality assurance documents submitted and the evaluation results.

* *Supplier inspection/audit/review*: the desk review record can be complemented with a checklist made during a plant visit. See regular audit working papers in Auditing Procedure section. Inspection is a specific inquiry as outlined on the desk review, while a review allows the supplier to participate in a qualifying action.

EXPLANATORY NOTE

The main purpose of this Procedure is that actual or potential suppliers be motivated to join in the quality assurance program of the organization and to participate in the selection, control, and overall qualifying process.

Categorization into major and minor supply assurance categories is related to the respective quality specification. Applicable product assurance standards, customer requirements, mandatory assurance requirements by regulatory bodies, etc., determine the form and content of the supplier qualification process.

Suppliers should be kept fully informed during the selection and performance control phases. In technical matters, suppliers often have more knowledge and experience, making cooperation mutually beneficial. Suppliers should become familiar with the organization's quality assurance policy and program and with their obligations as suppliers during the selection phase. When negotiating a contract, the Procedure for contract control is invoked and should have already been made known during the selection and qualifying process.

The quality assurance capacity for "just-in-time" systems must be assessed or possibly developed. The list of qualified suppliers should be distributed in a controlled and approved manner. Supplier quality assurance capability can be assessed by:

* the supplier's previous performance
* test samples
* receiving inspection
* audit of the supplier's quality assurance system
* satisfactory references
* recognized registration of the supplier's quality assurance system.

CONTRACT CONTROL

PURPOSE

The satisfactory performances of products and services, as specified in the design, production, and quality plans, are ensured by careful and systematic requisitioning, communicating, negotiating, and contracting.

POLICY

All supplies must be fully specified in purchase orders and other contractual documents, and quality assurance requirements and supplier obligations must be clearly stated, communicated, and accepted. Exceptions must be approved by senior management.

APPLICATION

This subsystem applies to all persons and departments authorized to specify supplies and subcontracts in conjunction with design, production, quality assurance, and requisiting of purchases, and who negotiate and monitor contracts with suppliers.

DEFINITIONS

A *supply contract* is an agreement between a qualified supplier and an authorized member or representative of the organization. In addition to the normal legal features of contracts, these supply contracts explicitly define quality and quality assurance obligations, including procedures to resolve disputes. Purchase orders and subcontracts are supply contracts. Drawings, descriptions of specifications, standards, and other written material related to obligations are part of the contractual documents.

Also see the definitions in Procedure for Supplier Relations.

REFERENCES

Applicable references are standards governing the items and services covered by the supply contract; applicable Quality System/Management Standards; directives of both customers and regulatory bodies; Procedures for Supplier Relations; Contract Control, Supply Verification.

STEPS/METHOD

1. Prepare a supply/purchase requisition that includes all the necessary specifications and quality assurance stipulations, references to standards, directives, associated procedures, and communications for purchasing agent and supplier. The requisition must be made by those responsible for production, use of supplies, and quality assurance.

2. Verify the clarity and completeness of the requisition pertaining to quality and quality assurance obligations. This should be done by supervisory management or the quality assurance department.

3. Add and communicate to requisitioning function any supplementary stipulations not included in the original requisition; these stipulations might include quality assurance tests or inspections to be performed and the results reported by the supplier; special inspection plans subject to approval; supply conditions concerning transportation, packaging, and storing.

4. Get supervisory or senior management's approval for offer to purchase for major contracts. Quality Assurance must participate.

5. Communicate offer to purchase to qualified and approved supplier. Determine the supplier's understanding, ability, and agreement to ensure quality conformance. Clarify the procedure to be used to resolve disputes or changes in supply contract.

6. Negotiate contract according to guidelines and stipulations from senior and supervisory management, monitor and verify quality assurance obligations, conduct in-plant reviews to ascertain the supplier's capability of meeting its contractual obligations, other than those already verified during general supplier qualification.

7. Finalize and implement contract and contract performance control. Establish schedule for source and receiving inspection, reporting, and decisionmaking instruction.

8. Communicate to the affected departments the supply performance and verify compliance with contractual obligations and procedures. When major deviations occur, conduct a review or audit, report the findings, and take corrective action. Control any changes in the contract; repeat as above.

9. After completion of the supply contract, complete supplier record and, if applicable, requalify the supplier for future contracts.

FORMS

The forms required by the subsystem include the following:

* *Purchase requisition*: This form identifies the item or service,

specifications, references to standards, last purchase data including the supplier, special documentation on quality assurance and tests, the department issuing the requisition, date, recommended supplier, associated order and purchasing documents.

* *Purchase offer/order*: This form identifies the item or service and specifications for quality assurance and test documents.

* *Supplier review*: This form provides item identification, references to the requisition, quality assurance requirements, checklists for review, purpose of review, results, and follow-up decisions.

* *Change notification to supplier*: This form identifies the change, the reason for the change, approval, identification of documents to be altered or withdrawn, date, price, and other implications. It also provides for verification and review.

EXPLANATORY NOTE

Supply contract control extends over the production and quality planning stages, requisitioning, and supplier selection to the actual control of contract negotiation, contract finalization, contract execution, and postcompletion activities, such as performance evaluation and recording. This Procedure should be applied in a discretionary manner: Relatively simple items or services with low cost and risk do not require a strict application of this Procedure, while complex and important supply contracts require additional technical and assurance stipulations to be added to the contract documents.

Supplies that are purchased frequently from suppliers with a satisfactory performance record and rating allow ordering to become a routine procedure with sampling controls and occasional requalification.

For new designs or production processes and new suppliers, this Procedure must be complemented by more careful and detailed communication and verification between supplier and purchaser. Whenever feasible and reasonable, a sample item or service should be obtained before final ordering or release of production in accordance with the supply contract.

In new or uncertain supply contracts or both, owing to still unproven supplier capacity or innovative design or process features, continuous communication and cooperation between purchaser and supplier must be arranged from the beginning of the contracting stages; technical personnel and quality assurance specialists of both purchaser and supplier must participate. A review will provide senior and supervisory management with proper control information.

Acceptance and receiving criteria must be agreed upon in a separate and complementary procedure, since this is an internal matter not subject to the contract itself.

Applicable methods for quality assurance of such contracts are:

* reliance on a supplier's quality assurance record or system or both,

* supplier inspection and test results,

* acceptance sampling by supplier or purchaser or both,

* purchaser source or receiving inspection or both.

An important prerequisite for any one or a combination of these control methods is that the supplier has received, understood, and accepted precise and consistent quality specification and quality assurance and inspection and test directives, and, if applicable, requirements for compliance with quality assurance system standards.

SUPPLY VERIFICATION

PURPOSE

All supplies of items and services when completed at a predetermined intermediary stage or at the final stage in accordance with the supply contract must be inspected for satisfactory compliance. This Procedure guides such a verification and should be referenced in major supply contracts.

POLICY

All supplies when received must be inspected in accordance with general and specific test procedures. The general procedure follows, while the specific procedure is subject to agreement with the supplier and should be part of the contract. Supervisory management together with Quality Assurance decide on corrective or preventive action. Supplier performance must be monitored and the results communicated to all interested personnel.

APPLICATION

This procedure applies to all persons and departments authorized to issue purchase requisitions and to negotiate and accept purchase contracts, and to those assigned to source and receiving inspection and to monitor supplier performance and the qualified supplier list.

DEFINITIONS

Source inspection is a formal, systematic, and predetermined independent verification that the product or process meets known and agreed-upon specifications. This verification is performed at the supplier's plant or an independent test laboratory, with or without a purchase representative being present.

Receiving inspection is normally performed on the end item at the time of delivery and at the purchaser's plant or production plant or both. Inspection and test procedures are to be determined, communicated, and applied in cooperation with the supplier and quality assurance department.

Corrective actions are determined and communicated by those persons or departments having released the purchase order/contract or their delegates. In major contracts this action and procedure must be prearranged and agreed to for proper and speedy resolution.

REFERENCES

Applicable references are standards for Quality Management Programs; Procedures for Contract Control and for Supplier Relations; Procedure for Acceptance Sampling.

STEPS/METHOD

1. Determine the need for source and receiving inspection for each major supply contract using production and quality plan, supply contract, and supplier record.

2. Decide on inspection point, object of and specifications for inspection, inspection or test method, inspector or team composition, and acceptance criteria. Prepare a formal inspection plan.

3. Attain approval for inspection plan; minor supplies might be inspected in a simplified manner with a routine or standardized procedure. Source inspection should be arranged and conducted in cooperation with supplier.

4. Arrange for planning and scheduling individual inspection or tests at the "source" or "receiving point."

5. Notify supplier of source inspection or receive notification of source inspection results, including any evidence gathered.

6. Prepare records of source inspection, by either supplier or purchaser, and identify inspection status on the item or any accompanying documents.

7. In cases of rejection, corrective action must be decided on, communicated to the supplier, and documented; refer also to Nonconformance Procedure.

8. Prepare for receiving inspection; appoint a person or department for minor and major supplies. Minor supplies might be inspected or tested according to a simple standardized procedure. Major supplies must be inspected in accordance with predetermined inspection or test procedures in compliance with contractual agreement. The inspection or test method depends on the complexity and importance of the supplies and performance record of the supplier. Standardized Acceptance Sampling Plans might be used; see Special Acceptance Sampling Procedure.

9. Rejection of supplies and corrective action will be conducted and recorded in accordance with Nonconformance Procedure. The supplier should be informed and participate in remedial and preventive action, and supplier's performance record should be completed.

10. Inspection or test results must be recorded, and inspected

items or lots marked with inspection status. Supplies should be stored under controlled conditions.

11. Changes in inspection plan for supplies and subcontracts must be approved by those issuing the original plan, and then this information should be disseminated. Any obsolete plan must be withdrawn.

12. Inspection performance should be supervised and periodically reviewed and audited. Inspection facilities and personnel must be conducive to valid and reliable inspection.

FORMS

Forms required for this subsystem include the following:

* *Supply verification/inspection plan:* This plan identifies the item or service supplier, supply contract, inspection points, methods, and inspector references.

* *Inspection tag:* The tag identifies the inspection status of the item.

* *Source inspection report:* This report identifies the item or service, inspection plan, inspection or test results, and documentation.

* *Receiving inspection log:* This log contains information on the item or lot or service received, references, inspection report identification, inspection result, corrective action, and dates.

* *Receiving inspection report:* This report, for major supplies, identifies item or lot or service, references detailed from inspection log, results, corrective action as taken, and dates.

EXPLANATORY NOTE

Source and receiving inspection and tests are crucial because supplier quality assurance is not directly controllable by the purchaser. The stipulation that for major supply contracts only qualified suppliers from the qualified supplier list should be used provides some assurance. Source and receiving inspection plans complement proper selection of and communication with suppliers in that this plan describes contract and product-related verification of contract compliance.

The Plan will be further operationalized through inspection and test procedures for the individual inspection points laid out in the plan. Less complex supplies with relatively little risk and need for quality assurance should be controlled through standardized and routine receiving inspection. Any inspection must remain cost-effective and therefore adapted to the real needs for verification. Suppliers with a satisfactory performance record justify less strict inspections. Standards for acceptance sampling such as MIL 105D have such flexibility in adapting to various needs and circumstances.

Nonconformance control requires special care because of the involvement of an outside supplier. Immediate rectification is necessary to prevent long delays and adverse impacts on the purchaser's ability to meet commitments. Rectification ranges from simple notification to cancellation of contract and claim for restitution.

Supplier performance records must show all major nonconformances and the reasons for them. When a serious supplier quality assurance deficiency occurs, either requalification as a supplier or withdrawal of qualification should be considered.

Defective supplies discovered at a later production or performance stage must be traceable when risk and cost warrant. Supply verifications, when based on clear, reliable, and justifiable procedures, induce control over preceding stages in the supply assurance, purchase contract control, and supplier qualification.

The supplier should know what receiving inspection method is to be applied so that any rejection can be assessed and rectified. Cooperation through exchange of test data and other related information is of mutual benefit and can result in agreement on ways of handling defective supplies and of preventing them, leading to general improvement of both quality and business relationships.

Modern computer-based and time-phased procurement and supply systems, such as "just in time," require reliance on downstream quality assurance and effective coordination of corrective measures in case of defective supplies. Receiving inspections, particularly when these are complex and time consuming, are not feasible and must be replaced by source inspections and system audits. When handled carefully, documents received with deliveries allow defective supplies to be traced to their source.

3.4 Production Assurance

Conformance to production and quality plans that are prepared for each product and service is the main task and objective of the production assurance subsystem. This internal "creation of products and services" needs adequate supplies, ensured through proper supply assurance procedures; facilities; equipment; and well-trained and -motivated staff. Given that the plans are well prepared, production start-up procedures aim at satisfactory work flow and workmanship. Other assurance procedures guide various inspections that verify and control the production of products or services and the staff and process performance. A procedure on self-inspection guides the quality inspection of the operator's own work. When major nonconformances occur, effective production assurance demands careful analysis and remedial action in order to prevent a recurrence of the "out-of-control" incidences. Procedures that ensure conformance to major quality specification normally contribute to the attainment of other customer and product or service specifications, such as meeting deadlines, provision for customer inspections, or required changes during production.

PRODUCTION START-UP

PURPOSES

The review of process/production capabilities. "first-piece" inspections, testing of machine and manpower performances, adequacy of supplies, and completeness and accuracy of plans and associated documents ensure adequate production performance and output. Adjustments and learning normally precede "steady-state" production, and this procedure is to facilitate and control these activities.

QUALITY MANAGEMENT SYSTEM

POLICY

Production readiness must be affirmed for each production run, with special care given for new products or services and for production or process conditions.

APPLICATION

All production supervisors and operators participating in production assurance activities are governed by this procedure.

DEFINITION

Production assurance comprises all activities that plan and control quality-effective production and conformance to specifications.

REFERENCES

References pertaining to this procedure include the following: system standards as applicable; product- or process-related standards, codes, and directives; referenced procedures, such as statistical process control; and individual test procedures.

STEPS/METHOD

1. Compile and review production and quality plans and associated documents. Complete documentation by acquiring missing information and directives.
2. Prepare inspection or test procedures if not yet included in the quality plan; assign, test, and adjust test facilities and equipment; train the inspector.
3. Conduct or review process capability studies and establish process control limits and criteria (see Procedure for Control Charts).
4. Verify availability of supplies for start-up run for all production stages. Start-up normally produces items for the entire production run.
5. Verify availability of all work and inspection instructions. Supervisors are to report on start-up readiness.

6. Conduct start-up briefing with Production Planning, Quality Assurance, plant supervisors, and inspectors.

7. Release start-up for first run/first piece and monitor performance, adjust the performance to conform to specifications.

8. Verify the adequacy of production based on start-up results and subsequent adjustments. Conduct special briefing for new products and processes.

9. For complex production assurance tasks, conduct a special quality audit whose results are reported to senior management and, if applicable, to the customer. This audit should also be conducted when final production approval must be obtained from senior management.

FORMS

The forms required by this subsystem include the following:

* *Checklist(s):* Checklists are needed to verify the product or service at each production or operation stage.

* *Approval for start-up and final production release:* This release identifies the product or service, the major results from the pre- and poststart-up reviews, and pinpoints weaknesses and areas requiring special attention.

EXPLANATORY NOTE

Depending on how thoroughly the production and quality plans were prepared and how actively involved the production staff was will determine how difficult the start-up and early implementation of these plans will be. While a "first-piece" inspection after each new setup or process control should always be performed, comprehensive start-up assurance procedures might not always be warranted.

Properly starting production processes leads to so-called controlled conditions with predictable outcomes. Factors that must be controlled are material; equipment; software; personnel; environmental conditions; and quality assurance, inspection, and test activities.

Process capability studies require statistical methods, which also facilitate statistical process control for subsequent operations and acceptance sampling. If process capabilities vary from the desired specifications and tolerances, capabilities have to be adjusted or the respective design standards must be assessed and revised.

All documents accompanying physical production provide important control data and evidence, once properly administered.

This Procedure should only be invoked for new products or services with relatively high risks and quality costs (inspection and defects), after major production changes, or when senior management and customers reserve to themselves final production release.

The start-up period offers ample opportunity for participation and cooperation in attaining a state of control. After any major start-up problems have been resolved, the start-up procedures including requalification of processes and operators might have to be repeated.

IN-PROCESS INSPECTION

PURPOSE

This subsystem provides for the uniform and effective verification of conformance to specifications concerning product, service, and processes at each predetermined point of the production flow and facilitates inspections by supervisors, operators, and customer representatives. This subsystem provides a basis for auditing inspector performance and for possible automated or computerized verification or both.

POLICY

All internal verification of conformance to specifications is planned and controlled so that nonconformances are properly detected, analyzed, corrected, and prevented for recurring. Wherever feasible, supervisors and operators should conduct the verification and perform corrective action.

APPLICATION

This subsystem applies to plant manager, supervisors, and quality assurance staff.

DEFINITIONS

A *process* is any planned sequence of activities or operations directed toward a desired and specified outcome under specified conditions. The process is in a controlled condition when output or performance criteria are within tolerance or control limits.

Inspection is an independent verification of conformance to a standard (specification) by a competent operator or inspector using approved methods. Inspection can be manual or automated or computerized.

Inspection point is the production stage where the verification is to be performed according to certain approved procedures.

REFERENCES

The references applicable to this subsystem are standards for Quality Assurance System, Production, and Quality Plans showing inspection points and general stipulations; individual inspection or test procedures for inspection points.

STEPS/METHOD

1. Assign inspection point and task to an independent, competent, and reliable inspector. Consider inspection by the operator, who is then responsible for the verification of conformance, performance, and workmanship.

2. Determine and provide data about the specifications to be verified, the method to be used, the criteria for acceptance or rejection, the appropriate action to take in cases of nonconformance, the records to be kept, the equipment to be used, and the special qualifications needed by the inspector or supervisor.

3. Document the inspection and test procedures and get approval from supervisory and quality assurance management and also from the customer's representative for predetermined inspection points that are subject to customer approval.

4. Verify adequate inspection performance; any adjustments are to be made before clearance by the inspector is given.

QUALITY MANAGEMENT SYSTEM 79

5. Supervise the inspection performance and the implementation of any adjustments. Support the inspector. Reduce or increase the frequency or scope of inspections when the production performance warrants.

6. Review inspection procedures after major nonconformances, production or process changes, and production and quality plan alterations.

FORMS

The forms used with this subsystems are as follows:

Inspection and test plans: These plans identify the product or service to be inspected or tested, inspection point, production run, time, etc.; inspector qualifications; inspection methods; equipment/facility; decision criteria; and nonconformance handling procedures.

Inspection log: This log identifies the product run, date, sample result, corrective action taken, and inspector.

Inspection tag: This tag lists the status of item or lot instructions, date, references, log, etc.

EXPLANATORY NOTE

Applicable standards for inspection normally do not permit self-inspection by the operator. Nevertheless, a combination of an operator and a special inspector might be acceptable and cost-effective. Obviously, inspection costs should not exceed the costs of defects that would be passed under normal in-control production conditions. Statistical process control methods are convenient and offer important advantages when they are combined with previously performed process capability studies. Moreover, statistical process control can be implemented together with electronic process control.

In situations where quality of the output from a process must be controlled indirectly through control of the process, such as brewing beer, control of the equipment and operation conditions becomes crucial. There is, however, no major difference in the verification principles, namely, checking conformance to specifications.

Similarly, workmanship can be controlled when specification and acceptance criteria are fair standards. Control of workmanship is best performed by the supervisor and operator. For crucial items and specifications, or after major changes that are dependent on workmanship, special inspections might be necessary.

Inspection points and objects inspected can include handling and storing of "work-in-progress." In large operations with complex and high-risk quality assurance, special inspection procedures for these "nonproduction" operations should be established. Here, again, inspection verifies specifications, or handling–storing facilities, or product-related specifications, including packaging. When inspecting services, the performance and immediate customer reaction and satisfaction are verified.

Special processes that have a critical impact on quality need to be controlled through special skills, equipment maintenance and operational supervision, software control, and operator- and inspector-planned and -supervised verification. Instructions for operations and out-of-control situations should be formalized, understood, and readily available.

If processes and process conditions are changed, the process start-up procedure normally must be repeated.

INSPECTION BY OPERATOR (SELF-INSPECTION)

PURPOSE

In self-inspection, meeting workmanship standards is verified by the operator through specially planned and supervised inspection and testing in order to

* make his or her work more meaningful (job enrichment),
* participate in quality assurance,
* reduce time between error detection and correction,
* utilize special expertise of the operator for cause analysis and error prevention,
* motivate suggestions for improvements,
* recognize good work performance,
* provide for career advancement.

POLICY

Quality assurance is an inherent responsibility and element in any job or work assignment. The job holder should, wherever it is feasible and has been approved, inspect his or her work. This inspection follows an approved procedure and an inspection or test method and is documented, the results being recorded and audited. One goal of self-inspection is to enrich the worker's assignment, but not necessarily to up-grade the job. If self-inspection is implemented, it must be economically and technically sound.

APPLICATION

The self-inspection subsystem applies to the quality assurance staff, supervisors, plant and production managers, personnel department, and operators.

DEFINITIONS

Self-inspection is where an operator inspects his or her own work in accordance with a predetermined inspection or test plan and other applicable procedures.

REFERENCES

The applicable references are Procedures for In-Process Inspection, Nonconformance Control, Quality Control and Inspection Resources, and Statistical Process Control.

STEPS/METHOD

1. The guiding principles of self-inspection are as follows:
    a. the worker must be trained and competent;
    b. the risk of defects passed has to be acceptable;
    c. the procedure and inspection or test method have to be approved.
    d. training and operational support, such as a statistical control chart, must be provided;
    e. inspection performance must be verified, supervised, and audited;

f. the self-inspection plan must be approved by senior management, quality assurance unit, worker representation (union), and customer (if required);
g. the implementation of self-inspection is to be planned and controlled by the quality assurance staff.

2. Discuss the self-inspection concept and its principles and application with all concerned personnel, i.e., supervisors, workers, and workers' representatives.

3. Form a project team (task force, Quality Circle) to plan, implement, coordinate, and supervise self-inspection.

4. Adopt the following steps:
    a. make a feasibility study and develop a project plan;
    b. obtain senior management's project and budget approval and any recommendations for changes;
    c. formulate a policy for job-integrated quality assurance and disseminate this policy to the staff;
    d. determine the kind of integrated quality assurance required, for example, simple end-item self-inspection, statistical process control, or comprehensive quality assurance from work design through completion;
    e. design and adopt a self-inspection procedure in the quality assurance system;
    f. review and change the supervisor's job description, and provide additional training if necessary;
    g. conduct a pilot study;
    h. analyze the outcome, modify the plan, and obtain senior management's approval;
    i. prepare detailed instructions for selection, training, support, supervision, and work place changes for jobs and operators with explicit quality assurance responsibilities;
    j. gradually phase–in the extended quality assurance system, with back-up controls (e.g., special inspection or audits);
    k. perform follow-up reviews.

FORMS

The forms required by this subsection include the following: application for self-inspection by worker, checklist for assessing its application, appointment form with self-inspection instruction, inspection or test method, control chart, observation form ("trouble report")

EXPLANATORY NOTE

Self-inspection is justified when the benefits for the worker, company, and customer are consistently attained, and quality assurance effectiveness and quality of products and services are maintained, if not improved. The main prerequisites for self-inspection are as follows:

1. The existence of a quality assurance system that is visibly supported by senior management and supervisors.

2. Relatively harmonious relationships among staff.

3. Supervisors' and quality inspectors' supportive attitude toward delegating responsibilities to operators.

4. A willingness among operators to accept quality assurance responsibilities in their assigned jobs.

5. Physical and social workplace conditions conducive to quality assurance by operators.

The operator should be provided with clear work instructions and a job description that includes details of the quality assurance assignment. The inspection plan that is prepared for the worker basically should be the same as the one prepared for quality inspectors; it could be simplified with control charts, acceptance sampling plans, and easily operated test equipment.

The sharing of quality assurance responsibilities relevant to the job in question can be described by a matrix as shown in Table 3-1. This describes four phases in each job, or work cycle, namely, the design and setting of performance standards, start-up, execution, and verification.

Statistical quality control charts are particularly useful for self-inspection. The operator can either conduct process capability studies independently or have the chart prepared either by the Quality Assurance expert or in Quality Circles. Owing to advancements in computer technology, operators, supervisors, and experts can develop computerized inspection instructions and tests when designing the job-related quality assurance.

Table 3-1. Batch Production, Self-Inspection Mode

Work Element	Operator	Supervisor	QA Staff	Planner
Design product, including testing	Inform	Inform	Prepare inspection	In-charge
Laboratory test	Inform	Inform	Testing	In-charge
Finalization	Inform	Participate	Approve	In-charge
Pilot run	Participate	In-charge	Approve	Approve
Set-up	Participate	In-charge	Review	Inform
Manufacture	In-charge	Supervise	Inspect	Inform
Process control	In-charge	Supervise	Inspect	None
Output inspection	In-charge	Review	Audit	Inform
Defect	Report	In-charge	Approve	Inform
Decision	Participate	Participate	In-charge	Participate
Follow-up	Participate	In-charge	Participate	Participate

Self-inspection basically formulizes and improves normal performance controls of a conscientious operator. Through self-inspection, the quality assurance staff and the supervisor help the operator meet assigned workmanship standards more consistently and reliably. Assigning self-inspection recognizes good work performance; it is not designed to prompt proper performance.

Self-inspection does not normally upgrade a job within formal job classification schemes and collective agreements. Extra time must be allowed for self-inspection and standards must be modified accordingly.

Craftpersons traditionally checked their own work. However, with the introduction of mass production, job specialization, and work standardization inspecting and testing were organized separately from the operator. Today, with modern test equipment and computer-based quality assurance methods self-inspection again has become possible. Supervisors and many workers are better prepared for inspecting their own work and having definite quality assurance responsibilities.

Prerequisites, as they have been listed, must be met for the successful introduction of self-inspection. Expanded work instructions include clear definitions of work assignment and delineations from related jobs. The quality assurance function coordinates self-inspection with other quality assurance activities.

Implementation must be carefully controlled, performance and developments must be monitored through supervision and audits, and improvements must be made when necessary and feasible.

END-ITEM INSPECTION

PURPOSE

The purpose of this subsystem is to ensure that the final product or service conforms with the design specifications and satisfies the customer and that all accessories and documents, such as customer information booklets, packaging, and test results are complete and accurate.

POLICY

All products and services must be inspected and verified before being released to customers. Coordination with the marketing and sales departments and customer services ensures quality of products and related services to customers.

APPLICATION

This subsystem applies to all supervisors, shippers, distributors, customer service staff, and assigned inspectors.

DEFINITIONS

End Item is any product that is complete as defined in the respective design and that is released for delivery by the production supervisor. For services, an end item is the rendering of the service to the customer, e.g., the completion of an insurance contract.

REFERENCES

References applicable to this subsystem include design assurance documents, design specifications, and respective standards; inspection and test instructions.

STEPS/METHOD

1. Compile design documentation and customer specifications (contract) concerning delivery, service, installation, and application. Review for completeness and accuracy.

2. Identify "final item," lot, and service, and have the production supervisor determine the release status.

3. Check traveling documents for completeness and correctness.

4. Prepare and retrieve the inspection and test instructions; acceptance sampling should use properly selected sampling plans according to Procedure for Acceptance Sampling. A representative unit should be tested, whenever feasible, under actual application conditions. The customer might assist in verification.

5. Conduct or supervise or both an inspection or test; if a rejection occurs, invoke the nonconformance procedure and take corrective action. Complete documentation.

6. Review and audit "end-item" inspection at adequate intervals and in accordance with the Audit Procedure.

7. Check customer-directed brochures and other software.

FORMS

The forms used with this subsystem are as follows:

End-item inspection and test instructions: These instructions identify the item, methods to be used, acceptance criteria, and nonconformance handling and procedure.

End-item inspection log: This log identifies the item, date, inspection and test results, corrective action taken, and signature.

End-item inspection tag: This tag identifies the item, references, traveling documents, instructions for handling, and inspection result and status.

EXPLANATORY NOTE

End-item inspection follows the same basic principles applied to any independent verification at the supply receiving point, during production, or at intermediate storage and handling. However, it is important, because the next "inspector" will usually be the customer.

For items that were produced for a market rather than for a known customer, an end-item inspector should cooperate with the marketing and sales department and with the distributors. Items not passing this final inspection point could be sold at discount when quality image and competition allow.

For items specified and ordered by a specific customer, the customer usually participates in nonconformance decisions if not in the end-item inspection itself.

For complex, high-risk items or services, the final inspection should not be performed by supervisors or operators; Marketing, Design, or Production might assist Quality Assurance.

For installation and initial service at the customer site, special quality assurance procedures should be established and applied; "end item" refers to the delivery stage at the plant.

A simplified end-item inspection is warranted for simple products or services, where this amounts only to verification and quality control. Such a procedure is often combined with the shipping function and can be arranged as a self-inspection by the supervisor or operator. (See Procedure for Self-Inspection.)

The verification status of inspected end items and related test results should be noted clearly on the item(s) and documents. This enhances traceability and clearly informs the shipper and customer of this information.

NONCONFORMANCE CONTROL

PURPOSE

Any detected deviation from specification (nonconformance) or any other defects must carefully and systematically be corrected, analyzed, and prevented from recurring.

POLICY

Anyone who observes a defect or nonconformance should report this immediately so that additional defects, expenses, and loss will be avoided and preventive measures and corrective action can be implemented. Nonconforming items must be tagged and safeguarded to allow for analysis of the problem, and corrective decisions and action to be made and taken.

APPLICATION

Since major nonconformances must be brought to the attention of all departments, this procedure applies to more people than the related procedure for inspection and testing. Any person internal or external to the organization participating directly or indirectly in controlling nonconformances, once these nonconformances occur, must apply this Procedure.

DEFINITIONS

Nonconformance is any deviation from specified quality characteristics in a product or service. Inspection and test procedures normally define major nonconformances in order to alert the inspector. In a procedure that describes and directs remedial and preventive activities, once a nonconformance has occurred, the definition of nonconformance should also include any perception of a deviation by other than official inspectors. Nonconformance is a more narrow concept than "defect."

Defect (defective) is any perceived fault, error, lack, failure, or other fact or occurrence that bears negatively on the usefulness of and expectations for a product and service. Defect(s) can render an item or service defective.

Corrective action is taken in accordance with proper procedures and authorization in order to remedy defects and nonconformances; when causes of defects are corrected, the need for future corrective action is reduced or even eliminated. Corrective action includes repair, rework, scrapping, or selling an item at a discount.

REFERENCES

References applicable to this subsystem are as follows: Quality Assurance System Standards, inspection-related procedures; technical, external product- or service-related standards, such as safety standards.

STEPS/METHOD

1. Identify nonconformance wherever it occurs; determine the severity of the nonconformance's impact on quality and safety and take corrective action. Immediate corrective action should be taken if it is authorized, it is inherent in the job responsibility, or severe risk and danger exist.

2. Clearly mark the item and any traveling documents that a nonconformance has been detected; nonconforming service must be stopped and withdrawn.

3. Segregate nonconforming items in a designated holding area, whenever practical and feasible.

4. Communicate and record details on nonconformance through a special report and communication channel; for major nonconformances and defects, a special alert procedure might be invoked.

5. The nonconforming item should be reviewed without delay by Quality Assurance and any other authorized person. Determine and undertake corrective action in order to prevent a negative impact by the nonconforming and defective item or service.

6. Inform all concerned personnel about the corrective action that has been taken as well as instructions on immediate precautionary measures that should be instituted.

7. For major nonconformances, so designated by the review in point 5, form a team or conduct a meeting with supervisory and quality assurance staff or both in order to analyze the causes and effects of the nonconformance. Prepare a report, including recommended preventive measures or other adjustments that are required.

8. Disseminate the report, request comments and suggestions, finalize preventive measures, and attain approval from supervisory or senior management.

9. Implement preventive measures and other modifications.
10. Verify the actual change and improvement resulting from the measures that were taken.

FORMS

The forms required by the subsystem include the following:

Nonconformance report/notification: This form identifies the item, defect, place, time, person who detected the nonconformance or defect, corrective action taken or recommended, references to related inspection procedures or job responsibilities or authorities. (This notification might be an oral report of "trouble" to the supervisor who then prepares a written report.)

Nonconformance tag: This tag identifies the item and defect, provides a warning in large letters, and identifies any action taken and the person or department to be contacted.

Nonconformance review report: This report provides for the transfer of data from nonconformance report notification and any additional data or records necessary for members of the review team, and identifies the members of the review team, the chairperson, the investigative and test measures taken and the results achieved, the corrective action recommended and taken, and the measures authorized and verified.

EXPLANATORY NOTE

Nonconformance correction and prevention are at the heart of any quality assurance system, whether it is a simple inspection system or a complex and comprehensive preventive nonconformance and quality assurance system.

Once the immediate problem has been remedied or brought under control, serious nonconformances have to be followed-up and recurrences must be prevented. Causes for "unquality" can rest in material, production, machine or operator performance, inspection testing, or among other factors that can be attributed to insufficient management involvement or inadequate procedures and quality assurance efforts and directives. The review team must be in a position to investigate and resolve the nonconformance and its contributing factors commensurately with the severity of impact and frequency of occurrence.

Nonconformance handling—including monitoring, communication, correction, and prevention—provides for meaningful participation in quality assurance. Only in exceptional cases should nonconformances, at least internally, be treated as confidential.

Nonconformance requires accurate and timely reporting and communication, which increasingly demands a computerized alert system and immediate correction, with subsequent analysis and prevention. Some nonconformance correction can and should be delegated to the department or person most familiar with the nonconformance and its cause.

Operators and other staff members should be alert for possible nonconformances and should be encouraged to report any occurrence or indications observed, and, therefore, no punishment should be involved with such reporting.

Occasionally, nonconformances are due to a justified deviation from obsolete specifications and procedures. The nonconformance review should take this opportunity for quality improvement.

QUALITY CONTROL AND INSPECTION RESOURCES

PURPOSE

This procedure ensures that all staff, equipment, and facilities used for any quality control activity render valid and reliable results.

POLICY

Inspectors must be properly trained, certified, and equipped. All test equipment and facilities must be in proper working order and controlled by supervisory management. Inspection can be assigned to the operator when reliability is ensured. Effectiveness and efficiency of quality control must be controlled.

APPLICATION

This procedure applies to all persons officially assigned quality control and inspection duties, and to those in charge of test designs, equipment, facilities, and follow-up, as well as their supervisors.

DEFINITIONS

Resources are human capabilities and material capacities available for quality control, inspection, testing, and decision making. Development and use of these resources generate costs (the inspection and prevention costs).

Calibration is checking and, if necessary, adjusting, test equipment against valid and approved standards and maintaining proper documentation.

REFERENCES

References pertaining to this plan include the following: inspection plans and procedures; Quality Systems standards; calibration standards.

STEPS/METHOD

1. Analyze inspection plans and procedures for resource requirements.
2. Check for resource availability and augment these resources, if necessary.
3. Assign an inspector, check his or her qualifications, and provide any necessary training.
4. Provide test equipment, check for its adequacy, and, if necessary, calibrate, replace, or repair it.
5. Determine inspection location; arrange and assign test facilities.
6. Prepare inspector job description and employment contract, including procedures for operator based inspection ("self-inspection").
7. Arrange for inspection supervision, resource management, and performance audits.

8. Safeguard and maintain equipment; replace the equipment when necessary.

9. Calibrate the equipment against a valid standard.

10. Properly dispose of obsolete equipment.

FORMS

The forms used with this subsystem are as follows: list of inspectors (including their qualifications, etc.); list of inspection and test equipment (including details such as calibration record, identification, and location); list of test facilities (including outside laboratories used); tags on equipment showing their maintenance records and current status.

EXPLANATORY NOTE

Resource management in quality control activities is influenced by such factors as costs of errors by this department, the confidence required in their decisions, and proof of proper care in case of product or service liability claims. Inspectors' activities and decisions are sometimes controversial, and any incompetence on the inspector's part can lead to disagreements. Quality control of all quality control and inspection activities should be arranged through training, requalification, and frequent audits.

When machine-integrated quality control is used, the equipment and software must be tested and maintained frequently. This is in addition to the regular equipment and facility maintenance.

Proper resource management requires an independent and sufficient budget, including resource accounting.

3.5 Product Performance and Customer Service Assurance

PRODUCT PERFORMANCE

PURPOSE

This procedure covers nonproduction and operation processes, such as storing, handling, and servicing, and seeks to avoid any defects caused in these areas or during these stages.

POLICY

All products in the care of the company or its agents must be properly stored, handled, and preserved in order to avoid any damage, spoilage, or other defects that impair product performance and customer safety. All services rendered on behalf of the company must be adequately planned, prepared, and rendered in order to ensure expected product performance and customer satisfaction.

APPLICATION

This procedure is applicable to staff, agents, and, to a certain extent, the customer or owner of the product; to persons in charge of these functions and those who might observe a violation of this policy; and to suppliers and subcontractors.

DEFINITIONS

Product includes material, parts, accessories, instructions, etc.

Product status is a record or label identifying the product's condition for use, e.g., inspected, not inspected, in-transit, condemned, or customer property.

Handling includes nonproduction or processing activities such as storing, transporting, displaying, assembling, or demonstrating.

Preservation includes packaging, labeling, maintaining (preventive and corrective), and staff or customer directives.

REFERENCES

References applicable to this subsystem include the following: procedures governing the previous listed activities; contracts with agents, suppliers, and customers; government regulations; laws; and standards for quality assurance. Procedure for nonconformance control.

STEPS/METHOD

1. Check production and service or operation plans for material flow and the previously listed activities and holding points.

2. Inspect each activity and location for quality assurance; make improvements if necessary.

3. Check for product safekeeping instructions or directives and the respective labels, tags, or traveling documents.

4. Check for identification of product or batch status labeling.

5. Check work instructions for operators and inspectors with regard to handling, storing, etc., instructions.

6. Check inspection and test plans for control of product safekeeping.

7. Check outside handling and servicing for explicit quality assurance directives, e.g., contract stipulations.

8. Analyze customer complaints, warranty claims, repairs, rework, etc., for those problems due to improper handling and servicing; make the necessary improvements.

9. Check packaging, containers, etc., for safekeeping of product and for enhancing performance and customer satisfaction.

FORMS

The forms required for this procedure include the following: special checklist for points mentioned previously; general defect and improvement report form; service/repair form.

EXPLANATORY NOTE

This procedure deals with many important and often hard to control causes for defects and failures. Human negligence is probably the main cause of problems in the activities and areas under consideration, because of the tendency to underestimate risks, the reliance on outside agents, and customer ignorance and, often, negligence. Quality of performance is too often taken for granted.

With modern computer-based material flow, procurement, handling, and distribution systems, such as "just-in-time", planning and control of these nonproduction and operation areas or stages have become crucial. Storage is reduced so that replacement of defective items beyond a minimal safety stock cannot be made in time. With the increasing technological complexity of modern products and services, care and maintenance by customers become more difficult, so that proper servicing by the producer or agent is critical to customer satisfaction.

Companies in service industries find that this section in their quality assurance system needs much more attention than this procedure allots. The term "operation" used, however, indicates our reference to pure service operations. This procedure can become the basis for more detailed quality assurance in pure service operations. "Pure" indicates that practically all companies provide services to customers, whether or not it is associated with a physical product. In a "pure" service operation the producer (server) and customer normally meet during "production" when service is rendered. Quality assurance of services to customers depends to a large extent on the qualification and personality of the server.

CUSTOMER RELATIONS

PURPOSE

This procedure ensures proper contact and communication with established customers of the company and its agents regarding product or service quality. It is also designed to create in the marketplace a positive quality image for the company and its products or services.

POLICY

All quality assurance efforts ultimately concern customer satisfaction and a positive quality image, which are conveyed through positive customer contact. Our customer must know about our quality assurance and must be able to contribute to satisfactory product or service performance. Any customer complaint or suggestion needs to be handled with the utmost care.

APPLICATION

This procedure applies to staff in contact with customers, customer representatives, and publication media dealing with quality assurance measures; to all other staff and agents because they contribute to the general quality image of the company; and to agents providing services on behalf of the company.

DEFINITIONS

Customer includes potential customers of the company and its agents who are addressed through advertisements, brochures, etc., or who are in negotiation with the company. This procedure is concerned with any customer interest in quality (including safety, information) and quality assurance.

Customer relations include any communication or personal contact on quality and quality assurance matters.

REFERENCES

References needed for this procedure include the following: any instructions or procedures dealing with customer relations; standards such as ANSI/ASQC Q94-1987. Procedures for contract control and nonconformance control.

STEPS/METHOD

1. Prepare and issue a code of ethics (behavior) to staff members in regular contact with customers or who are responsible for customer service; conduct training and orientation, if necessary; or establish a procedure.

2. Check customer complaint and suggestion handling for quality assurance; assess the procedure and the results.

3. Analyze customer feedback to identify where improvements in customer information, product support, servicing, and general quality image are needed.

4. Check customer-directed pamphlets, advertisements, brochures, manuals, and other publications with regard to quality, quality assurance, and general quality image.

5. Conduct customer surveys to determine quality and quality assurance perception and quality image; analyze the results and make improvements.

FORMS

Forms used with this procedure include the following: Code of Ethics and General Quality Assurance information pamphlets; general defect improvement notification form; customer complaint and suggestion record; brief customer questionnaire.

EXPLANATORY NOTE

This procedure might be considered as dealing with some commonsense matters. But proper customer contact helps to resolve serious dissatisfaction, helps to prevent it, and can create a harmonious business relationship. Quality and quality assurance are always major matters of concern, not only when a customer is dissatisfied and complains. The surveys mentioned inform current and potential customers of the company's concern about and efforts toward improving quality.

Positive comments by customers and a good public quality image foster satisfaction and pride among staff and company agents, which also reinforces quality of the work and general quality assurance.

This procedure concentrates on quality and quality assurance, and is related to many other procedures that deal with customer and customer relations; therefore, overlaps and conflicts might exist.

Proper customer relations are of short-term importance in "pure" service industries where direct interpersonal contacts dominate.

A close relationship with distributors and customers ensures open and timely feedback and notification of design defects and nonconformances. Inspection by distributors and customers through the active life of a product could be arranged to obtain useful data for modified or new designs. This could also extend to a competitor's designs and design changes.

SERVICE QUALITY ASSURANCE

PURPOSE

This procedure ensures satisfactory product installation, performance, warranty services, repairs, and other product-related services to customers.

QUALITY MANAGEMENT SYSTEM

POLICY

The quality of any service rendered to customers for ensuring the proper performance of a product within contractual and reasonable limits must be planned and controlled.

APPLICATION

This procedure applies to any staff member who is assigned customer service duties, the supervisor in-charge, and agents contracted for such service.

DEFINITIONS

Service in this Procedure is rendered upon a customer's request or with a customer's approval to ensure, enhance, or restore proper product performance.

REFERENCES

References applicable to this Procedure include service contracts; standards and guidelines, e.g., ANSI/ASQC Q94-87; and the other procedures in this Guide. Procedure for nonconformance control.

STEPS/METHOD

1. Identify product- and product-performance-related customer services in contracts; design specifications, production plans, and assembly or installation plans.
2. Inspect the company's and agent's resources for ensuring the quality of service.
3. Check service and work instructions for quality assurance directives.
4. Check for inspection procedures of service quality assurance.
5. Analyze customer complaints or suggestions for quality of service improvements.

6. Check for enhancing customer self-service and improved quality of service.

FORMS

Forms used with this procedure include the following: service order forms with quality assurance instructions; service inspection form; and general quality assurance improvement observation form.

EXPLANATORY NOTE

Even when the warranty period has elapsed, customers request and need various services and support for product performance. The product identifies the company, and poor repair or information service negatively affects the company's quality image. When customers trust in the product-related expertise and integrity of a company's service and the company's backing of its products; they will more likely accept advice for replacing the item when it is beyond repair.

The availability of spare parts and the updating of information for distributor and customer ensure proper service and also maintain meaningful contacts.

Customer service is often rendered by agents, who are not all under the company's control. This Procedure attempts to guide such service agents for the mutual benefit of the company and the agent.

3.6 Quality Management Information System

Each of the preceding procedures is a documented subsystem and is available as information, which directs and supports quality assurance. Forms that are associated with these procedures induce and facilitate data compilation. Data processing and communication provide information to those concerned; alert the reponsible parties in case of major defects; initiate corrective action; and facilitate managerial review, planning, and control. The Quality Management Information System must be designed as a special subsystem of the Quality System and is integrated with the general companywide information system.

The following procedures are closely related to all other quality assurance procedures, and stipulate a quality report as a regular or ad hoc management information source. The procedure for quality cost accounting is designed to facilitate managerial control of the economics and the long-term profitability of quality assurance. These Procedures are not concerned with the explicit quality assurance of information and the management information system.

QUALITY REPORTING

PURPOSE

Quality reports summarize quality assurance developments over the short to medium term, and include statistics on defects and quality costs, major problems, and improvements effected. Ad hoc reports, which augment regular reporting, induce decisions and actions without delay.

POLICY

Regular and ad hoc quality reports keep management, supervisors, and other designated persons informed about quality assurance matters. Additional reports should be provided upon request or when warranted. Reports designed as public information should not contain any confidential information. Reports and underlying reporting should foster open internal communication on all matters of quality and quality assurance. In addition, only processed and analyzed data should be communicated to facilitate effective and convenient comprehension and to elicit the desired reaction.

APPLICATION

This procedure applies to receivers of reports as designated by senior management; to persons responsible for data compilation, reporting, and report preparation; and to all staff members through prescribed forms and approved or open reporting and communication channels.

DEFINITIONS

Report is a formal or informal regular or ad hoc written communication that contains useful information. Report formats can be prescribed, but in quality assurance matters it is important that all major defects and major occurrences are reported. Quality reports, therefore, range from a simple defect observation report to the immediate supervisor to a comprehensive quality report for senior management. In a hierarchical and well-structured reporting scheme only important and useful reports are prepared and communicated.

Reporting is a communication along designated channels on quality assurance matters by any person concerned or when important observations concerning quality and quality assurance have been made.

REFERENCES

Applicable references are company procedures for information and communication systems; forms associated with the preceding procedures; quality system standards.

STEPS/METHOD

1. Determine management's regular information needs concerning quality and quality assurance developments and issues.

2. Assess, categorize, and coordinate these information needs in conjunction with preparing a tentative report format and reporting schedule.

3. Determine data availability, reporting source, format, and communication channel and mode.

4. Appoint a report coordinator who has sufficient authority and support.

5. Establish or approve reporting scheme and schedule, and test these.

6. Survey or audit receivers of quality reports for the information's usefulness.

7. Augment regular quality report and reporting with an accessible data bank on quality assurance data.

8. Establish ad hoc quality reports and reporting scheme for defect alerts, product recalls, and information that cannot be delayed.

9. Create a data bank and establish authorized access to it.

FORMS

Applicable forms include quality report; special form for report data and other contributions; intermediary and partial quality reports; ad hoc quality reports; "trouble" report form.

EXPLANATORY NOTE

Quality Information System design and operation is a comprehensive and complex task. Reliable and valid information drives the system. The quality report(s) must be perceived as useful information and should increase attention to and interest in quality matters. Everyone actively and passively involved should benefit from the information and the reporting. Reporting should be nonpunitive, constructive, convenient, timely, factual, and concise.

While confidential matters need to be controlled, receiving reports and contributing to them might extend to suppliers, customers, and the general public; for example, a newsletter could enhance regular reports.

Quality reports should measure up to quality standards of format and content. They should meet actual needs for information and available sources for respective data, and be standardized for convenient and easy comprehension and decision making. Modern computer-based reporting and communication add to effective and timely reporting.

A quality indicator condenses current data. An example is $QI =$ (percentage of defectives or defects) (weight factor) $(e^{-0.1})$ (100), which is used in the HILTI Corporation. Relative costs of defects are considered by the components of this Indicator. QI is compiled for different time periods, plants, departments, and products and is compared with respective costs for defects and defect detection and prevention. Extrapolation of these data allows for forecasting. Statistical control charts can also be applied.

QUALITY COST ACCOUNTING

PURPOSE

Quality cost accounting determines the impact of quality assurance in terms of cost, cost reduction, and contribution to profit. It supports planning and control of all decisions and activities to ensure the economic effectiveness of quality assurance.

POLICY

Costs incurred through quality assurance activities and through defective quality are to be compiled, analyzed, and reported at regular intervals and on an ad hoc basis in order to optimize total costs, to support quality assurance and corporate management, and to make monetary benefits of quality assurance visible. Quality cost accounting is integrated with general cost accounting, and general accounting standards apply.

APPLICATION

This procedure applies to staff assigned to recording and reporting costs related to quality assurance, to accounting staff, to information systems staff, and to receivers of cost reports.

DEFINITIONS

Quality cost is the monetary value of resources expended for ensuring quality, and attributed to defective quality, of products and services. It is the sum of prevention, appraisal, and failure costs. Higher prevention costs should reduce failure costs, and, in the long run, quality costs.

Prevention costs, such as for conducting audits, are those costs incurred to prevent defects and failures.

Appraisal costs are costs for inspecting and testing supplied products or services, work in process, end items, processes, returns, etc.

Both, prevention and appraisal costs are quality assurance costs compared with defective and failure costs.

Failure costs are caused by company or plant internal and external failures, such as avoidable scrap, rework, repair, or returns.

Other cost classifications and compilations, such as fixed or variable, actual expenditure or opportunity cost, are applicable in quality cost accounting.

REFERENCES

Applicable references are Quality Assurance System Standards and various publications on quality costs by the American Society for Quality Control, Milwaukee, WI.

STEPS/METHOD

1. Clarify the concept and method of quality cost accounting in general.

2. List cost examples for each of the three cost categories—prevention, appraisal, and failure—and keep the list open for additional entries.

3. Determine existing quality cost data for those cost elements and categories listed.

4. Prepare detailed and comprehensive cost recording and reporting forms and obtain approval for them. Some costs are recorded on existing forms, such as for nonconformance control.

5. On a preliminary basis, compile costs for the most recent period, analyze these by means of cost ratios, and submit them to senior management or project committee.

6. Modify preliminary cost recording or reporting as required, complete gaps, determine cost reporting with regard to cost breakdown (by period, product, product line, plant, units). Determine who is to receive cost reports and what the respective report's content and format will be.

7. Discuss preliminary quality cost accounting system with designated contributors and users, delineate cost elements and provide for changes and additions, determine the need for

cost information and the application, safeguard confidentiality and reliability, and assign responsibilities.

8. Finalize quality cost accounting system with accounting and information system and integrate with management cost accounting.

9. Standardize recording and reporting forms and procedures and provide any necessary training.

10. Start up system and ensure feedback for necessary modifications.

11. Audit quality cost accounting for compliance and effectiveness.

12. When it has been implemented successfully, enhance the system by adding the costs of quality improvement projects.

13. Apply quality assurance methods for quality cost accounting, e.g., cost control charts, data recording reliability analysis, cost forecasting and tracking, and quality-oriented audits.

EXPLANATORY NOTE

The actual need for cost information drives the cost recording and reporting system. This information assists in planning and controlling ongoing quality assurance decisions and activities; major changes and improvements in the system; special quality improvement projects; and new-design quality assurance costing, budgeting, and performance assessment.

Principles of constructive cooperation beyond functional and departmental boundaries prevail. System integration serves this purpose as well as economizing the effort. Close cooperation with Accounting is particularly important.

Quality cost accounting lends itself to computerized recording and reporting or communication schemes, rapid updating, standardized data processing and report preparation, and convenient retrieval of cost data and information.

Definition and delineation of quality costs in detail are often difficult and allocations must be made by estimates. Confusion and conflicts can be avoided through prudent and nonpunitive use of cost information.

The main purpose and emphasis of quality cost accounting is to reveal potential improvements of the quality assurance system, to justify investment and expenditures, to make profitability of quality assurance visible, and to establish realistic quality objectives.

3.7 Statistical Quality Planning and Control

Statistical methods applied to quality assurance include

* Statistical process control

* Acceptance sampling

* Design of experiments (Taguchi Method)

* Analysis of variances

* Reliability analysis

In this Guide we provide Procedures only for the first two methods, since they are essential elements in any quality assurance system.

The two basic methods are statistical process control by means of various control charts and acceptance sampling for finished-lot evaluation. Both sampling methods can be applied to measurements and nonmeasurable quality attributes. Sampling offers low-cost and reliable quality control that is scientifically sound and widely applied and recognized.

Experimental design and the related Taguchi Method should be applied when the impact of different variables on quality is to be studied. Target values can be attained that are technically and economically sound.

STATISTICAL PROCESS CONTROL

PURPOSE

Unnatural variations in a process should be detected, and assignable causes for the nonrandom variations should be corrected. The process capability is to be confirmed by comparison of desired and actual ("natural") statistical control limits and random variation of plots.

POLICY

As an integral part of the quality assurance program and applicable quality and inspection plans, statistical sampling must be applied in a cost-effective, reliable, and valid manner to establish statistical control limits and to control process performance. The technical means of achieving these goals are control charts for variables and attributes.

APPLICATION

This procedure applies where the quality produced can be quantified, and the lot size and production run permit sampling from the ongoing process, including test runs and set-up procedures. The information derived from the applicable control chart must clearly permit maintenance and output quality improvement, as expressed in major and critical measurements or attributes. The supervisor and operator are guided by this procedure in conjunction with expert advice.

DEFINITIONS

Statistical method obtains information from numbers.

A *statistical control chart* shows control limits and a centerline; plots from samples indicate the state of process control.

A *process capability study* samples data from a process, determines the control chart criteria (centerline, upper and lower control limits) normally for ±3 Standard Deviations, and compares these with the desired values. Deviations measure relative process capability and allow for adjustments, such as process improvement or change of desired values (tolerances).

REFERENCES

For the statistical techniques any book in the field can be used; the best books of course are those on statistical quality control. For instance, Besterfield, Dale H., *Quality Control,* Prentice-Hall Inc. 1979 or Sinha, M., Willborn, W., *The Management of Quality Assurance,* John Wiley & Sons, Inc., 1985.

Canadian Standards Association Special Publication Z90-1975. American National Standard ANSI/ASQC A1-1978, *Definitions, Symbols, Formulas, and Tables for Control Charts,* and *Standards Handbook 3, Statistical Methods,* of International Standards Organization (ISO), Technical Committee 69.

STEPS/METHOD

1. Determine the objectives of and benefits gained from the application of statistical process control for the product, process, inspection point, or quality characteristic under consideration. Some objectives are reduction of waste, process improvement, performance control, job enrichment. Some benefits are cost savings, more reliable quality control, higher productivity, and meaningful participation in quality assurance.

2. Communicate with supervisors and operators about the application of statistical process control and discuss further improvements of specific controls. Provide training and practice in the use of control charts; demonstrate the effectiveness of control charts.

3. Determine for a specific process the quality characteristic, phase, time, place in the process that best assesses and expresses performance concerning quality.

4. Select the proper control chart(s); for example, X-bar and R charts provide more detailed information through variables, while a p-chart expresses general overall performance with larger sample size and can also be applied to attributes.

5. Conduct a process capability study to attain conformance to specifications. Special procedures might have to be established: determination of desired limits, sample size, etc.; collect data from a process after that process is checked when it is being performed under normal and representative conditions; determine confidence level and respective control

limits; plot data on chart; interpret pattern of plots; compare desired and actual limits; and decide on process capability.

6. Decide on the format of the control chart and prepare it; instruct the operator or control chart user.

7. Have the operator, with expert assistance, establish the centerline and the control limits; have the plots analyzed to assess the status of current process performance. Decide with the operator on the validity and usefulness of chart.

8. Determine time (frequency) of data collection (sampling) and recording after calculating the plots.

9. Supervise the proper application and use of the chart; allow time for learning; review user interpretation and decisions.

10. When process out-of-control condition is indicated, follow an established procedure: immediate action and communication, assessment of required analysis of cause, corrective action, verification that assignable cause has been corrected.

11. Reset or review chart and process status when major changes have occurred, e.g., new material, process change, different operator.

12. Audit the need, benefit, and effectiveness of the control chart; report this audit and have supervisory management decide the issue.

In some practical situations a simplified procedure can achieve similar reliable controls.

SIMPLIFIED PROCEDURE

Step 1: Determine quality characteristic on drawings, etc., for assessing process capability; include nominal and tolerance values.

Step 2: Select an adequate control chart.

Step 3: Enter desired centerline and control limits; use predetermined number of Standard Deviations or Calculation Factors.

Step 4: Take samples from ongoing or pilot production or process; ensure the validity and reliability of samples.

Step 5: Calculate centerline and control limits; compare these with those of Step 3 and decide on process capability.

Step 6: When desired process capability is confirmed, use control chart to control the process. Repeat Steps 1–5 when changes are made.

FORMS

The applicable forms include control chart forms for various kinds of process control; standard charts are available from the American Society for Quality Control, Milwaukee, WI.

EXPLANATORY NOTE

Acceptance sampling can control quality of completed lots and thus can be used for additional process control. See Procedure on Acceptance Sampling. The advantage of control charts compared with acceptance sampling is the earlier detection of both nonconformances and deviations from expected random variation. Information derived from each control chart must always lead to continued good production and productivity. This principle demands frequent review of control chart design and application. Changes are necessary when the information does not serve properly for work planning, control, and quality-oriented decision making. Changes include variation in frequency of sampling, selection of new quality characteristics, process control points, inspector, type of control chart, and confidence limits. The customary ± 3 Standard Deviation statistical control limits correspond to 99.7% Confidence Limits for the Normal Distribution.

Once simple control charts, the X-bar/R chart and the p chart have been established for effective process control, other types of control charts could further refine the statistical process control. Standards, which were referenced in the preceding Procedure, explain such advanced control charts. Users of control charts can gain valuable information and guidance from them without understanding the underlying statistical theory, although such knowledge would render this technique more meaningful and interesting.

The control chart method should only be applied when the information is actually required and process performances are analyzed. When the process seems well in control with regard to one quality criterion, sampling frequency can be reduced, other criteria selected, or other measures taken in order to minimize inspection costs.

Control charts lend themselves well to operators inspecting their own performance, e.g., self-inspection.

ACCEPTANCE SAMPLING

PURPOSE

This procedure guides the acceptance, or rejection, of a lot (batch, finite statistical population) on the basis of testing a representative random sample and predetermined statistical properties and decision criteria. Reducing inspection costs without compromising the reliability of a decision is also a goal.

POLICY

Acceptance sampling should be used for inspection and testing when its results are reliable and valid, the method has been agreed upon by all parties, applicable standards are used, and cost-effectiveness is attained. Average outgoing quality (AOQ) should be applied.

APPLICATION

This procedure is applicable to all persons in charge of inspection and test planning, to suppliers, and to customers, where prerequisites can be attained and sampling risks are acceptable.

DEFINITIONS

An acceptance sampling plan states sample size(s) and associated acceptance number (e.g., allowed defectives) for a given lot size.

QUALITY MANAGEMENT SYSTEM

Acceptable quality level (AQL) is the maximum percentage level of defectives, or proportion of variant units, for good quality.

Lot tolerance percentage defective (LTPD) is the level of poor quality with low probability of acceptance.

Statistical risks are those risks of rejecting a good lot (alpha risk) and of accepting a poor lot (beta risk).

Average outgoing quality (level) (AOQL) is the expected outgoing quality after an acceptance sampling plan has been used.

REFERENCES

Applicable references include standards for accceptance sampling, e.g., MIL-STD-105D, MIL-Std-414, ANSI/ASQC A2-1978, ANSI/ASQC Q94-1987; Sinha, M., Willborn, W., *The Management of Quality Assurance,* John Wiley & Sons, Inc., 1985; *Standards Handbook 3, Statistical Methods,* International Standards Organization (ISO) Technical Committee 69.

STEPS/METHOD

1. Survey applicability of acceptance sampling to supply assurance, production, and delivery.
2. Determine who will be involved for each case; negotiate an agreement among all affected personnel and outside parties.
3. Determine acceptance sampling standard or system, e.g., MIL-STD-105D.
4. Determine acceptance sampling plan(s), including AQL, LTPD, risks, single or multiple sampling.
5. Incorporate acceptance sampling into the respective inspection or test procedure.
6. Instruct the inspector or testor, and ensure the reliability of their decisions.
7. Authorize the use of acceptance sampling and inform the parties involved.
8. Conduct inspection or testing using acceptance sampling and keep records.
9. Review or audit the application of acceptance sampling.

FORMS

The forms needed for this procedure include inspection or test plan with incorporated acceptance sampling plan; inspection log; inspection or test record.

EXPLANATORY NOTE

Acceptance Sampling can be used externally with suppliers and customers and internally at inspection points. Its application should always be statistically sound and confirmed by an expert. Use of standards, as previously mentioned, has considerable advantages. These standards are generally recognized, e.g., MIL-STD-105D, and are flexible.

Additional details on acceptance sampling are available in the literature, such as Dodge/Romig *Sampling Tables* and textbooks on statistical quality control. For practical applications the underlying statistical theory and the so-called "OC" curves need not be fully understood by the user when standards are applied.

The concept of "acceptable quality level" and respective tolerance for defectives have led to some misunderstanding. Acceptance sampling plan on per mille rather than per cent basis have been published recently. Moreover, in computer-based quality assurance systems, 100% inspection has replaced acceptance sampling to a certain extent.

Acceptance sampling can be conducted for attribute and variable testing, and can also be used in conjunction with statistical process control using control charts. Both methods complement each other, one controlling the ongoing process and the other the final or intermediate output (lot).

Acceptance sampling is also applicable in service industries.

For every acceptance sampling plan the "average outgoing quality" (AOQ) curve can be determined showing the outgoing percentage of defectives in relation to the incoming percentage of defectives. With inspection and replacement of defectives made, an "average outgoing quality limit" (AOQL) in terms of percentage of defectives can be promised to the customer.

3.8 Software Quality Control

With modern computer-assisted production and operations, management software quality and quality control have become essential elements of any quality assurance system. Most generic standards describing quality assurance systems have not as yet incorporated procedures for controlling software design, implementation, application, and maintenance. There are, however, an increasing number of additional standards and guidelines that deal with the subject. A generic quality assurance system without procedures for software quality assurance must be considered incomplete.

The following is a general procedure for software quality control. This could be augmented through more detailed technical directives in each case of application.

SOFTWARE QUALITY CONTROL

PURPOSE

Any software used must consistently meet its intended purpose and function. This procedure is to ensure the attainment of this goal. Because of the relatively new and unfamiliar subject matter in conjunction with quality assurance activities, this procedure is a preliminary one. Suggestions for improvement should be submitted to management responsible for the quality assurance system.

POLICY

After thorough research and justification, this organization is to use any suitable and applicable computer software, and respective hardware, to make its management and operations more effective. To ensure achieving the most benefits from such computerization, software quality must be carefully planned and controlled. Major principles of this software quality control are error prevention starting with acquisition, contracting, design, and continuing through implementation, application, maintenance, and phasing-out. Those responsible for the application of the software and quality assurance staff must participate at all stages. Standardized software packages and subcontracting should be used as much as possible in order to optimize the implementation, quality, costs, and servicing.

APPLICATION

This procedure applies to all persons requisitioning, planning, purchasing, producing, implementing, and using software and to quality assurance system managers and auditors.

DEFINITIONS

Software encompasses all instructions and data inputs to a computer for electronic processing, and the associated procedures, documentations, manuals, and operating systems.

Software quality is the degree of intended fitness for use and actual user satisfaction. Quality characteristics are the software's validity and reliability; its ability to be transferred, changed, adapted, and utilized for various applications; to be integrated with other systems laterally and hierarchically; and to be standardized. The software should be free of errors and deficiencies by virtue of the design and planning of the application. Latent and actual errors should be relatively easy to detect and to correct.

Software quality control comprises all decisions and activities that ensure software quality from the user's point of view. Planning, since it sets performance specifications and standards, precedes control. Individual quality controls are integrated and documented in a cohesively structured software quality control program. This program should be part of a comprehensive quality assurance program or system.

REFERENCES

Major standards for software quality assurance programs are ANSI/IEEE 730-1981 *Standard for Software Quality Assurance Plans;* NATO AQAP-13 *Software Quality Control System Requirements;* NATO AQAP-14 *Guide for the Evaluation of a Contractor's Software Quality Control System for Compliance;* and the Canadian Standards Association Q396.1 *Software Quality Assurance Program.*

General (generic) standards and procedures for quality assurance systems apply as well.

STEPS/METHOD

1. Establish a steering committee for software quality control. This committee is to set objectives for software quality control, and to initiate, direct, and supervise the planning and control or audit of a special software quality control program. Form project teams, if necessary.
2. Determine current software needs and use. Prepare a list that shows title, date, producer, applicant, and application or quality record.
3. Conduct a review of the software inventory with regard to quality and quality assurance. Report the results to senior management.
4. Initiate and prepare a Software Quality Assurance Program that complies with one of the preceding standards and includes and documents the following:
    a. Purpose
    b. Reference to standards
    c. Responsibilities and organizations or project teams
    d. Documentation and configuration control
    e. Identification of standards, practices, and conventions for documentation, logic structure, coding, and commentary
    f. Project planning and control
    g. Contract review
    h. Design specification review
    i. Supplier and contractor control
    j. Code and coding control
    k. Testing and error-correction control
    l. End-design review, verification, and validation
    m. Installation and implementation review
    n. Error and nonconformance control
    o. Audits of software quality control program and projects
    p. Change and modification control

5. For each software project and application the steps are:
   a. Person requisitioning the software determines and provides full documentation including intended application, current respective data and information/decision system/practice, intended application of software, expected improvements through computerization, general design specification, available software or supply sources or both, laterally and hierarchically related software, quality assurance and test plans or requirements, software maintenance, cost and other related financial data, staff and user qualification for software application and maintenance, hardware requirements, and due dates. Technical support for preparing the requisition should be provided.
   b. Steering Committee appoints reviewer or project team for assessment of the requisition with report to Steering Committee.
   c. Upon approval, the project team is confirmed for controlling software development and implementation. A special software quality control plan, possibly as part of a project plan, is prepared and submitted.
   d. Upon approval, person requisitioning the software and the intended user will be informed and will participate in formulating an inspection or test plan. This plan specifies inspection points and mandatory hold points during the development phases, decision criteria, test methods and environment, responsible person or department, equipment, budget, nonconformance or error correction, and due dates.
   e. Supplier or producer of software or both are selected and informed about quality specifications, along with other specifications and contractual obligations.
   f. Upon receiving a supply bid, person responsible for quality assurance verifies supplier capability and communicates and validates the software inspection or test plan. Any changes of the plan must be approved.
   g. Contractor prepares own inspection or test plan, including those for subcontractor(s), and submits this for approval and integration with user's plan.
   h. Inspection or test plan will be operationalized through in-

QUALITY MANAGEMENT SYSTEM 119

dividual inspection or test directives and resource deployment and will then be submitted for approval to person in charge of the inspection or test plan. The inspection or test capability is verified.

i. The inspection or test plan is coordinated and integrated into the general project plan. Modifications are subject to approval.

j. Inspections or tests are carried out according to plan and procedures with results recorded. Nonconformances are reported without delay, and remedial actions are carried out according to special procedures.

k. Upon completion, the software to be delivered is inspected for verification of contractual specifications. The project manager submits the results with all documentation to the Steering Committee for approval.

l. Upon delivery, all inspection or test records along with the software and associated documentation and manuals are submitted to the user for a formal, joint design review. The project team conducts or initiates and supervises this application-oriented software design review.

m. Upon clearance after the design review, the software is released to the user for controlled preliminary implementation. Inspection and testing are performed according to approved plan and procedure. Results of the inspection or test are reported to project team.

n. The project team releases the software for preliminary and conditional application and ensures that any deficiency or nonconformance is properly reported and corrected. The user informs the project team when the software should and can be fully approved.

o. The project team, upon the user's request for final clearance and approval, verifies through audit the satisfactory application and software maintenance methods and capabilities. The project team reports to the Steering Committee and requests to be disbanded.

p. The Steering Committee verifies software quality, based on documentation and possibly on a special software quality audit. After final clearance of software and software quality control the software application is listed and incorporated in regular audit programs. The project team is disbanded and the user is authorized to proceed with the software's implementation.

6. The Steering Committee oversees all software development, maintenance, and phasing-out or substitution projects.

7. The Steering Committee supervises software acquisition and application through quality control audits. Additional surveys and studies ensure that computerization and respective software are economically and technologically sound in design and application.

8. The Steering Committee follows up the development of standards and guidelines for software quality control and submits the software quality control plan to external audits. The program should comply with a standard included in preceding reference.

9. Quality Assurance Management incorporates software quality assurance, requiring the Steering Committee's full cooperation and coordination of activities. Audits are an important method for supporting and ensuring optimal integration, compliance with applicable standards, and effective maintenance and improvements.

FORMS

The forms applicable to this procedure include Software Quality Control Program/System Manual; software inventory, requisitions, project plan; change control; inspection or test plans and procedures; nonconformance handling; design review; and audit plans and checklists. Manuals and forms should be integrated with the principal quality assurance system.

EXPLANATORY NOTE

The preceding procedure can be simplified as well as extended and augmented through individual procedures. A simplification is recommended when the computerization and respective software are relatively unimportant for the entire operation. In this case, however, the potential benefits from further computerization should be assessed and the advantage of an early, careful phasing-in of modern technology by means of software quality control should be considered.

The simplified procedure would combine the Steering Committee and the individual project committees. Moreover, general quality improvement projects are the proper vehicle for the assessment and development of company and user-oriented software. The standards in the preceding reference direct any simple or complex software quality control.

In a situation with advanced and complex computerization in operation and administration, the Steering Committee will normally develop and implement more specialized control procedures. These would involve hardware and firmware acquisitions and productions, organizational arrangements and directives, security and safety procedures, and advanced research and development projects designed to foster the competitive strength through well-researched, developed, and dynamically applied and controlled software and related human and physical resources.

Decisions on software quality control are supported by various reviews and audits. Audits are an independent and formal examination and should apply respective auditing standards and guidelines. The AQAP-14 guide for evaluation of software quality control systems is an excellent support and includes checklist questions.

Proper software quality control is the basis for any computer-assisted quality control of products and services. With the introduction of software quality control, other quality control procedures and methods can be introduced, if they do not yet exist. Because of the serious repercussions and costs of computer errors, quality control of software is often more readily accepted than controls of manual processes and outputs. Software quality control can act as a leverage for a more comprehensive quality control program. Software quality control, and the respective program, should never be isolated from the overall quality control in an organization.

The user, like any customer, should become involved in all phases of software development and implementation for the obvious reason of error prevention. A major cause of error and deficiencies is misunderstanding and misjudgment of applications or changes in application. Our procedure observes this principle of user participation and continuous orientation to the intended application of software.

Other specialized software quality control procedures might have to be prepared as the application expands and becomes more technically complex.

4
Implementing the Quality Management System

When implementing a new or revised quality management system, starting points and ending points are not quite clear. Does the system start during the design phase? Does the implementation ever end when we should consider quality improvement as an unending task? Without doubt, however, only a well-implemented system can achieve its intended purpose.

We consider implementation of the quality management system as a crucial phase that needs to be planned and controlled properly. The following general guidelines assist in preparing

1. an implementation plan,

2. a quality manual,

3. workshops.

The three projects can be carried out concurrently or sequentially. The implementation plan launches the quality assurance program; the quality manual provides the essential information and documentation about the system; and workshops introduce the program to the staff.

As was stated previously, implementation happens not only once but quite frequently. With every major change in the system or in individual procedures, perhaps as a result of a review or an audit, the implementation guideline should help. The quality manual will then have to be revised too, and workshops will provide for the necessary updating and training.

AN IMPLEMENTATION PLAN

PURPOSE

The implementation plan is designed

* to assist in the effective introduction of a new or revised quality assurance program or individual procedures or both;

* to ensure consistency, acceptability, and understanding of the process;

* to foster cooperation and to motivate personnel to work for improved quality assurance.

POLICY

The implementation of a new or revised quality assurance program, in whole or in part, is a project subject to planning and control. Staff members, who are confronted with changes in their workplace and their performance standards, should actively participate in bringing about the intended improvements. Every necessary assistance should be provided. New procedures and work instructions designed for quality and quality assurance improvements should not be merely imposed, they should be carefully phased-in.

DEFINITIONS AND PRINCIPLES

Implementation is a process and project for introducing a new system into the mainstream of a company and the individual workplaces and jobs. The effectiveness of the implementation depends essentially on the soundness of the system's or program's design, the ability of the people involved, and the management's competence. Implementation is an intermediary phase that overlaps with design before implementation and with the subsequent system maintenance.

The following are some implementation principles:

* Adapt implementation to the existing situation and personalities involved; the implementation should be as informal as possible.

* Avoid disturbing ongoing work and operations as much as possible, with special meetings and time-off being scheduled; involve managers, supervisors, and staff members constructively during the designing phase.

* Test the program or procedure in the field to verify any improvements that have been assumed.

* Provide well-prepared documentation, information, and training.

* Facilitate compliance through well-designed forms and records, and provide supervisory assistance.

* Introduce the program or procedure on a preliminary basis and provide for feedback and modifications.

* Launch the program or procedure officially; follow-up on performance.

* Recognize outstanding contributions and excellent performances.

REFERENCES

Prerequisites for successful implementation are documented and approved procedures and manuals that describe the program. Consistent reference to these official information sources aids smooth implementation. Prototype procedures and other guidelines should be referred to, even when such a procedure has been modified. Applicable standards and guidelines, as referred to in the individual procedures and manuals, should be available during implementation.

SCOPE AND APPLICATION

Depending on the extent of and difficulties encountered, the planning and control of implementation will vary. Once started, the process of implementation should be adapted to the actual progress and problems encountered. Some implementation can be simplified and the time span shortened and vice versa.

Decisions and activities for smooth, early, and successful implementation of the entire Quality Assurance Program are addressed to the company staff. This should precede the introduction of procedures in respective individual units and workplaces. The type and extent of changes and improvements determine the people who become involved in implementation.

APPROACH/METHODS

1. Determine the approach to implementation before or while designing the new system or procedure.
2. Allow for early participation in design projects and activities by all concerned people.
3. Compile relevant documentation and aids for introducing the new program or procedure into the company and workplace.
4. Check program for consistency with applicable standards.
5. Test individual procedures in the field; assess difficulties; determine necessary support (apply special test procedure to be discussed subsequently).
6. Get approval for program or procedure and any resulting changes.
7. Launch program; conduct meetings about it; publicize it; and inform the workers.
8. Assign responsibilities; coordinate the project; provide support; and correct problems and difficulties.
9. Monitor the progress made.
10. Finalize the program or procedure.
11. Prepare and conduct audits and reviews.

Testing of implementation of individual procedure:

1. Identify the procedure to be tested, and get approval for its implementation.
2. Contact and inform the supervisor and operators affected.
3. Explain and discuss the procedure, using standards, guidelines, and other suitable and relevant aids.

4. Select and train person(s) to test and verify validity and suitability of procedure in the workplace.
5. Determine testing method and prepare checklists and other records.
6. Arrange "real-life" normal conditions at the workplace for testing.
7. Conduct the test and record observations.
8. Analyze and communicate results.
9. Provide for additional information, training, and support as necessary.
10. Introduce procedure, perhaps tentatively, and monitor its performance and the compliance with it.
11. Respond to feedback; review and modify the procedure when necessary.
12. Conduct a formal audit, and recognize acceptance, cooperation, and performance.
13. Get final approval.
14. Insert procedure into the Quality Manual.

TECHNICAL AIDS

* Prototype procedures in this Guide.
* Checklists.
* D. Hutchins, *Quality Circles Handbook,* Milwaukee, WI, American Society for Quality Control, 1985.

QUALITY MANUAL

A Quality Manual describes the quality assurance of a company and serves as an important information source during the program implementation. The manual should always be prepared because it communicates senior management's policy clearly, explains the organization, and includes all procedures and references to work instructions.

When individual procedures are introduced separately and the program is phased-in, the procedures in the manual are at first preliminary. Only when the entire program is implemented is the Quality Manual also complete as a formal document.

Preparing a quality manual is normally a new and unique task, so we provide a guideline for its preparation. In large organizations it should be handled as a project; in smaller organizations a simpler approach and manual format might suffice.

PURPOSE

The purpose of a Quality Manual is

* to document the Quality Assurance Program and its individual procedures,
* to inform about and guide its implementation,
* to facilitate training and auditing,
* to serve as a reference.

POLICY

The Quality Assurance Program is described and documented in a Quality Manual. This manual must reflect actual practices and compliance with applicable standards clearly and truthfully. Preparation of and revisions to the manual must be controlled carefully, and a distribution list should show all persons who receive it. The manual should be readily accessible as a reference and guideline. The manual and its distribution list must be reviewed regularly, at least once a year.

DEFINITIONS AND PRINCIPLES

A *quality manual* describes the quality assurance program of a company, states the policies to be followed, lists reference standards, shows the quality assurance department and its organization, and includes all quality assurance procedures. In a comprehensive quality assurance program several interrelated manuals are prepared. The corporate manual entails policies and general directives, while each organizational unit has its own manual. Detailed technical work instructions are related to the respective quality assurance procedure. A quality manual differs from a quality assurance plan, which describes all quality assurance activities with regard to a product, service, or contract.

The following principles result in a valid and useful quality manual:

* The manual must describe the quality assurance program as it is actually implemented.
* The manual's format and content must comply with standards and guidelines.
* The manual's format is standardized and consistently applied.
* The manual's content is complete and displayed for convenient reading and reference.
* The manual's language, style, and illustrations aid comprehension by a nonspecialist.
* Confidential matters are not included but referenced.
* Detailed technical procedures should only be included for major processes; these could also be referenced.
* Detailed quality, test, and inspection plans should not be included.
* Management uses the manual to convey its commitment to the program.
* The manual is readily available, and its distribution is controlled mainly for facilitating updating and renewals.
* The manual could be used for such purposes as publicizing the program and for training.

The manual's format is not stipulated in program standards. A loose-leaf binder allows for easy exchange of updated procedures, copying, circulation of individual procedures, correction, and replacement. A computer file might also be prepared in addition to the manual.

Each page is formated as follows:

* Company name, organizational unit, logo
* Manual title
* Blocks in heading for
  — procedure title and status (e.g., "preliminary")
  — procedure number
  — last revision date
  — page number and total number of pages
* Blocks for signatures, comments, cross references

Three major parts are

1. general program description including policy statement and quality assurance organization,

2. procedures,

3. appendix for forms, records, stamps, etc., used.

The manual's contents is predetermined by the quality program standard applied and the respective scope of the program. Table 4-1 is an example and includes the procedures in this Guide.

REFERENCES

Program standards and guides include brief statements of purpose. Major customers and program assessment and registration agencies have checklists for preaudit assessment of the manual.

APPROACH/METHODS

1. Appoint a coordinator and assign responsibilities; possibly prepare a matrix showing manual element (e.g., procedure) and those who are participating and what their responsibilities are.

2. Determine scope, format, and content of the manual.

Table 4-1. Quality Manual—Table of Contents

1. General
   a. Policy Statement and Certification
   b. Amendments
   c. Foreword
   d. Organization
2. Quality Assurance Procedures
   a. Design Assurance and Contract Review
      i. Design Assurance
      ii. Drawing and Specification Change Control
      iii. Design Review
      iv. Contract Review
      v. Process Capability Studies
   b. Quality or Inspection Plan
   c. Supply Assurance
      i. Supplier Selection and Performance Control
      ii. Purchasing and Subcontracting
      iii. Source and Receiving Inspection
      iv. Supplier Audits
   d. Production Assurance
      i. Production Readiness and Start-Up
      ii. Work-in Progress Inspection
      iii. Special Process
      iv. End-Item Inspection
      v. Nonconformance Control
   e. Quality Control and Inspection Resources
   f. Physical Product Handling and Care
   g. Product Performance Control and Customer Service
   h. Quality Management Information System
      i. Quality Report and Reporting
      ii. Statistical Quality Control
      iii. Software Quality Control
   i. Quality Assurance Audits
   j. Quality Improvement Projects
3. Appendix, Forms, etc.

3. Check with stipulation in program standard and guide; make modifications to the planned manual.

4. Conduct a meeting for orientation and workshops on procedure writing. (See our Guideline for Procedure Writing.)

5. Prepare general part of manual, e.g., policy statement and organization.

6. Initiate and supervise procedure writing, set deadlines, edit drafts, get approval.

7. Incorporate procedures into the manual as they are approved.

8. Collect forms and records, check these with the respective procedure including cross references and incorporate them in the appendix.

9. Determine distribution, registering, and revision procedures.

10. Once the manual is complete, get approval for issuing it.

11. Prepare and issue submanuals, e.g., supplier manuals.

12. Audit the manual and its implementation and invite an external audit of the manual.

13. Keep manual updated and revise or reissue when necessary; withdraw and destroy obsolete documents.

The following is a simplified method:

1. Prepare working papers for procedure writing.

2. Select procedures from this Guide and adapt them.

3. Check procedures with existing practices and modify either procedure or practices.

4. Enter procedures into a loose-leaf binder.

5. Complete general part and appendix of the manual.

6. Have the manual reviewed; modify it when necessary.

7. Distribute manual and keep record of holders.

8. Review manual regularly.

TECHNICAL AIDS

For the manual, the prototype procedures and checklists in this Guide should be used. A manual is outlined in the Appendix. Also see Carlsen, Gerber, McHugh, *Manual of Quality Assurance Procedures and Forms,* American Society for Quality Control, Milwaukee, WI, 1981.

GUIDELINE FOR QUALITY ASSURANCE WORKSHOPS

Implementing the quality assurance system and its procedures normally requires training. Among many educational and training arrangements, this subsystem concentrates on introducing quality assurance into workplaces. With changes and additions in practices and jobs, normally some training will enable the staff members to adapt and comply. Workshops will be conducted for groups and then followed up by individual instruction and self-learning modes.

PURPOSE

This guideline is designed

* to ensure proper planning, execution, and follow-up of workshops;

* to integrate these workshops with other educational and training programs;

* to simplify the task for coordinator, auditor, and instructor;

* to facilitate effective implementation of new or revised system and procedures.

POLICY

Training in the form of workshops will be arranged when quality assurance responsibilities are assigned and immediate compliance cannot be expected. The quality assurance coordinator and supervisors must monitor the need for workshops and ensure their effectiveness, including their cost-effectiveness.

DEFINITIONS AND PRINCIPLES

In a *workshop* participants learn technical subject matter in a practical and hands-on way. They acquire knowledge, skill, and initial experience for adequate performance and compliance with procedures and standards. The instructor is and remains cognizant of the participants' workplace and job requirements and guides and assists the participants to meet these.

Principles for these quality assurance workshops are:

* Ability of staff members to comply with new or revised quality assurance principles should be assessed and specific training needs clarified.
* Workshop format, content, timing, place, and instructor must be conducive to the training requirements.
* Additional preparatory and subsequent individual training should be arranged as required in individual cases.
* The workshop should inform, encourage, and motivate staff members in matters of quality assurance and its importance and relation to their own work experience.
* The composition of groups should be conducive to learning and personal enrichment.
* While the workshop is being given, it should be tested or adapted or both to the prevailing circumstances and needs of the organization, and it should be followed up.
* The instructor's qualifications and the value of the workshop should be verified.
* Adequate resources and learning aids should be available.
* Learning should be measured through tests.
* Formal recognition and certification should be considered.

The types of workshops are

* internal or external,
* general or technically specific,
* for management, supervisors, or workers, or any combination of these staff members.

REFERENCES

Applicable references are

* Guidelines and prototype procedures in this Guide,
* *Guide for Section Conducted Courses* (*noncredit*), American Society for Quality Control, Milwaukee, WI.

SCOPE AND APPLICATION

This Guideline applies to workshops that are implementing a quality assurance system and procedures; however, it is general enough to be applied in other workshops. It does not lead to a more comprehensive educational program, but is an ad hoc arrangement for very specific needs and circumstances. It might be applied in conjunction with other structured shop meetings and quality circle activities.

Quality assurance specialists, coordinators, consultants, instructors, supervisors, suppliers, and participants are addressed by these guidelines.

APPROACH/METHODS

1. Determine the need for workshop training:
   a. Survey the organization's current technical competence with regard to quality assurance responsibilities, writing of procedures, and understanding of quality assurance concepts and techniques.
   b. Delineate and profile the participant groups.
   c. Invite applications for training; assess suitability of workshop mode.
   d. Audit current quality assurance practices and implementation of quality assurance procedures; identify training needs.

2. Prepare workshop plan, which entails:
   * title that indicates type and subject matter
   * survey and assessment of need
   * purpose and objectives, general and specific
   * participants profile and training need
   * enrollment and registration
   * major content (see Technical Aid section in this Guideline)
   * mode of instruction/training/learning
   * preparation for workshop by participants
   * instructional and learning aids and materials
   * instructor name and qualification

- duration, time, and location
- assignments, testing, and certification
- follow-up services
- workshop evaluation and checklist
- recognition
- sponsoring and funding
- administration

3. Send plan to instructor and supervisors for recommendations and to senior management for approval.
4. Prepare final schedule, course material, etc.
5. Publicize, accept registration, determine classes or groups, send notices to participants.
6. Secure funding and contracts.
7. Launch workshop; invite representative of senior management.
8. Assign preworkshop activities to participants.
9. Conduct workshop according to plan; make changes that improve learning.
10. Administer tests and workshop evaluation.
11. Assess plan and workshop experience and evaluation; modify these for repeat workshops.
12. Follow-up workshop, and arrange for additional on-the-job training and other individual support.
13. Report outcome of workshop to senior management and supervisors; retain plan and reports for future reference.
14. Arrange for recognition and certification.

A simplified procedure is as follows:

1. Determine the need for workshop training and identify its objective, content, and mode.
2. Secure funds, instructor, and classroom.
3. Prepare a plan.

4. Announce workshop.

5. Register participants.

6. Conduct workshop.

7. Follow-up workshop and provide additional instruction.

TECHNICAL AIDS

* *Guide to Section Conducted Courses (noncredit)*, American Society for Quality Control, Milwaukee, WI.

* Contact local ASQC section and/or community college for assistance.

* Check ASQC catalog for suitable materials, e.g., standards, guides.

* See audit checklists in this Guide.

Some examples of workshop topics include the following:

* Guidelines and procedures in this Guide.
* Quality assurance concepts and policy:
  — General introduction in this Guide (Chapter 1)
  — Quality assurance and quality of products and services
  — Quality awareness, commitment, and action plan
  — Total quality assurance and other principles
  — Benefits from improved quality and quality assurance
  — Quality costs and quality assurance infrastructure
  — Quality assurance aspects of the industry
  — Quality assurance organization of the company, individual responsibilities, workmanship standards
  — Quality assurance program of the company and standards
  — Quality manual and individual documentation
* Quality assurance procedures and guidelines for implementation (Use applicable prototype of this Guide or modified procedures or both.
* Quality improvement concept, need, and technique.
* Statistical quality control methods and techniques.

* Problem solving using statistical techniques.
* Planning and arranging quality circles.
* Testing and inspection planning and control, preparation of quality/inspection/test plans.
* Quality assurance for suppliers.
* Quality assurance for distributors.
* Quality assurance in design, research, and development.
* Quality assurance and marketing (personnel, accounting, etc.).
* Quality assurance information system, recording, indices, reports, software, etc.
* Auditing of quality assurance program.

5
Auditing The Quality Assurance System

Audits help to determine when the system is adequately implemented and a "steady state" has been attained. Any system or individual quality assurance procedure that as an "open system" is influenced by factors and changes internal or external to the company needs to be examined and reviewed regularly. Management can monitor degeneration and weaknesses in the quality assurance program through observations, reports, and, in particular, audits.

In large organizations with comprehensive quality assurance systems and quality improvement projects auditing is formalized. In small enterprises audits are mainly undertaken by their major customers. Whether auditor or auditee, both must understand auditing principles and methods for mutual benefit.

The following Guideline is designed to facilitate sound auditing and to augment published auditing standards and official guidelines. In the subsequent special section we present audit checklists and summarize, with questions, the major aspects of establishing and maintaining a quality assurance system.

AUDIT GUIDELINE

PURPOSE

This Guideline is to assist

1. managers who initiate audits and receive the report;
2. auditors who plan and perform audits;

3. auditees who provide information and assist the auditor;

4. experts who advice auditors in technical matters.

The purpose of an audit is:

* to examine the effectiveness of the quality assurance system, of compliance with standards and procedures, and of quality improvement projects;

* to serve management as an independent information source and control device in making decisions;

* to examine supplier quality assurance;

* to provide for open communication on quality and quality assurance matters;

* to confirm that the system is properly maintained and identify further improvements that could be made;

* to prepare for external audits;

* to communicate management's concern and commitment to quality and quality assurance.

POLICY

Management controls and maintains the Quality Assurance System through special audits that are in addition to regular quality reporting and performance controls. These audits are initiated and approved by senior management and conducted in accordance with applicable auditing standards.

APPLICATION

This auditing guideline applies to senior management; to the department being audited; and to auditors, auditees, and other participants in audits. It can also be used for self-audits, which are those performed by the auditee in order to examine quality assurance.

DEFINITIONS AND PRINCIPLES

Quality assurance audit or *quality audit* are a formal and systematic evaluation of the Quality Assurance System, especially the system's compliance with applicable standards and its effective design and implementation. The object of the audit is predetermined and can be the entire quality assurance system, individual procedures, product-based systems, contracts, processes, or plants. Auditing is described in special standards and guidelines. An audit differs from inspection, survey, and review, although they are similar activities.

An auditor is the person, or team, qualified and appointed for an audit project. Auditors are often organized in an auditing unit. An auditor for an audit project is normally appointed by senior management that initiates the audit and receives the audit report.

Audit function/unit programs audit projects, employs auditors, and administers the audits.

Auditee is the person, or unit, in charge of the object to be audited and sharing the responsibility for a successful audit.

Auditing standards are published documents that describe authoritatively all important features of audits and auditing. These standards ensure the quality of audits and the resulting audit reports. See example of these standards under Technical Aids in this Guideline.

Some auditing principles are as follows:

* management initiates the audit, determines its objective, appoints the auditor(s), provides resources, and receives the report;
* the auditors comply with applicable and approved audit standards, assess their qualifications for the specific audit project, and maintain their independence;
* an auditor is not acting as a consultant with regard to the specific audit project;
* preaudit activity assesses the existence of a documented quality assurance program;
* the audit is planned and checklists are prepared;
* the auditor examines the evidence before judging compliance to standards and system effectiveness;
* observations are recorded with sufficient factual and clear evidence and are communicated to the auditee;
* auditee contributes to the success of the audit;

- the audit is used as a learning experience and as a means for communication;
- the audit report is widely distributed in order to enhance the audit's effectiveness;
- necessary confidentiality is observed;
- management and auditee take corrective action, which a reaudit might verify;
- audits must be cost-effective and of proven value.

REFERENCES

Applicable references include the following:

- Auditing Standards, e.g., ANSI/ASQC Standard Q1-1986 *Generic Guidelines for Auditing Quality Systems*; Canadian National Standard CSA Q 395, *Quality Audits*.
- Auditor certification Program, American Society for Quality Control, Milwaukee, WI.

STEPS/METHOD

1. Senior management initiates an audit, after a need for an audit is determined and approved. Supervisors can request an audit of their unit. Audits might be initiated when an audit program that includes various audit projects is approved.
3. An auditor, or audit team, is appointed. Qualifications for the auditor(s) is to be established.
3. Applicable audit standards are determined and assessed.
4. Audit objectives and objects are determined and documented.
5. An audit plan is prepared and is augmented by audit checklists and other forms and working papers. The auditee might participate in the audit's planning and preparation.
6. The audit is announced and conducted in accordance with

the audit plan and applicable auditing standards. Meetings are held to inform and guide auditors and auditees about observations made and further audit proceedings.

7. When an observation is made on a noncompliance or a potential improvement, all details are recorded, so that decisions can be made and corrective action taken.

8. Upon completion of the audit, the principal results are formally reported to the senior management that initiated the audit. Auditees should be kept informed about the report's content and observations.

9. Senior management decides on corrective action and might require a follow-up audit.

10. An auditor might conduct a follow-up audit. After final completion of the audit project, the audit plan, all working papers, and the audit report are filed for safekeeping and future reference.

11. If the audit team or auditor is not employed by an auditing unit, a proper release from other duties should be made by senior management. Good auditing performance could and should be recognized.

A quality audit approach for a small business is as follows:

1. Current quality assurance practices are determined. A "practice" is what the staff actually does to inspect and ensure quality. Good practices will be maintained.

2. Practices are recorded in writing. This formulation is brief, factual, understandable, and accepted by the person concerned.

3. The manager or owner approves, or disapproves, the documented practices. The manager or owner has the authority for this decision, not the auditor.

4. The approved, and possibly modified, practices are edited and integrated into the daily routine. Inconsistencies with overriding quality assurance policy and between individual practices are adjusted. All practices now become official

"tentative procedures" and are entered into a preliminary Quality Manual.

5. The tentative quality assurance system, as documented in the Quality Manual, is adjusted for compliance with the applicable Quality Assurance System Standard. The Standard is to be used by auditor and manager to ensure the adequacy of the system and acceptance by customers and regulatory bodies.

6. The documented and standardized system is augmented by forms for recording and reporting (information subsystem) and by a department to be responsible for the system. These are important prerequisites for the implementation, the control of performance, and the analysis and future prevention of defects. The Quality Manual is to be completed accordingly.

7. The new Quality Assurance System is implemented effectively into the mainstream and ongoing operations of the company. This critical step changes informal practices into formalized procedures. Some training and support will be required.

8. An audit system and service are organized for the maintenance and improvement of the Quality Assurance System. This is to ensure that the System will be examined at regular intervals and adapted to changing conditions. Under certain conditions the manager or other supervisory staff can conduct the audits.

9. The successful implementation of the System is audited and verified; this is to prove the usefulness of the audits and to open communication channels among management and staff. This audit is also to prepare for external audits.

10. The system is to pass an external audit and is officially certified and possibly registered. A major customer will be invited to conduct the audit and approve the System. Some further adjustments are normally required. Standards and Quality Manual serve as the major documents for this and any other future audit.

After working through these ten steps in close cooperation, the manager and auditor should remain in contact. The auditor has helped the manager to establish the System without taking on any responsibility for its operation. The auditor, therefore, can conduct internal audits in accordance with acknowledged auditing standards.

FORMS

Forms applicable to this subsystem are Audit Project Plan, Audit Observation Form, Audit Checklist(s), Audit announcement notice, audit report.

TECHNICAL AIDS

Some technical aids to refer to include

* standards, guidelines, 'how to' publications by the American Society for Quality Control, Milwaukee, WI.;
* auditor training courses and certifications;
* published checklists (e.g., see next section in this Guide);
* statistical tables for auditors;
* services provided by external auditors and program registration organizations.

EXPLANATORY NOTE

Auditing of quality assurance is an excellent learning and training opportunity for managers. Only larger companies with an extensive quality assurance system can afford a special auditing unit; otherwise, the auditor might be appointed on an ad hoc basis. When the audit plan and working papers are well prepared and standardized, the audit execution is simplified. Then, any manager, or supervisor, who is sufficiently knowledgeable and objective can serve as an auditor. The supervisor in charge of the operation to be audited might even conduct the audit, once the supervisor's objectivity is ensured.

Audits are designed to make improvements, and only to a lesser degree to evaluate individual performances. When "auditing without fear" is accomplished, audits enhance communication and motivation.

A prerequisite for sound auditing is the use of, and compliance with, published auditing standards. An example is the ANSI/ASQC Q1-1986, "Guidelines for Auditing Quality Systems." The 'Q1' Standard is compact. Individual clauses are illustrated in a flowchart so that one can get an overview of the Standard. There is a selected bibliography and a listing of other related auditing standards included. The document is relatively inexpensive and is sold by the American Society for Quality Control, Milwaukee, WI.

The audit standard, being generic, does not differentiate between product audit, process audit, or system audit, because there are no basic differences. According to the Standard, the objective of the audit project defines the audit object. (See Table 5-1.)

An important prerequisite for an audit is the existence of a documented quality assurance system. This Guide should have helped to accomplish prerequisites for passing audits. Moreover, the Quality Manual can now be submitted to any external auditor as a preliminary proof for the existence of the system subject to the audit.

Auditing involves statistical sampling in order to gather valid and reliable evidence. The auditor can practice, therefore, statistical quality assurance methods.

5.1 Audit Checklists

The following are some questions that could be included on checklists. Possible responses, to be checked-off in a box, are acceptable, not acceptable, and not applicable. These auditing aids can be augmented by identifying the evidence to be examined and by steps in the auditing process. The auditor must adapt the checklists in an audit and might add or delete questions. Completed and signed by the auditor, these checklists are retained as official audit document.

Sometimes auditors provide copies of the checklists to auditees, so that they can conduct their own examination (a "self-audit"). In our Guide we also use these questions to review our Guidelines and Procedures. Managing also means to ask the right questions, at the

Table 5-1. Generic Guidelines for Audits of Quality Systems, Content

Contents

1. General Features of Audit Standards
   1.1 Purpose
   1.2 Scope
   1.3 Premises and Concepts
   1.4 Applicability
   1.5 Definitions
   1.6 Audit Systems/Programs
   1.7 Audit Quality Assurance
2. Auditor
   2.1 Responsibilities
   2.2 Qualification
   2.3 Independence
   2.4 Performance
3. Auditing Organizations
   3.1 Auditing Teams
   3.2 Auditing Department/Group
4. Client and Auditee
5. Auditing
   5.1 Audit Initiation
   5.2 Audit Planning
   5.3 Audit Implementation
6. Audit Report
   6.1 Drafting
   6.2 Content
   6.3 Review and Distribution
7. Audit Completion
   7.1 Follow-up
   7.2 Record Retention
8. Quality Assurance of Audits

right time, to the right people. We hope that this Guide has made your managing of quality assurance more pleasant and successful.

GENERAL QUALITY ASSURANCE MANAGEMENT

1. Are managers and supervisors aware of and committed to quality assurance (QA)?

2. Are managers aware and knowledgeable about recent trends in quality assurance?

3. Do managers raise questions about quality assurance when interviewed? (See our questions in the Introduction.)
4. Are QA responsibilities assigned to them?
5. Is there a written QA policy?
   a. Is the QA policy consistent with product and standard requirements?
   b. Is the QA policy adequately communicated and known?
   c. Is the QA policy reviewed periodically?
6. Are managers aware of QA program standards and Guidelines?
7. Is the standard appropriate?
8. Do managers compile quality cost?
   a. Do they know and apply quality cost concepts?
   b. Do they communicate and analyze quality cost?
   c. Are quality costs compiled for internal and external failures?
   d. Are defect prevention costs compiled?
   e. Are appraisal costs compiled?
   f. Are quality cost indices and ratios compiled?
   g. Are quality cost data retained and retrievable?
   h. Is quality cost accounting integrated with general cost accounting?
   i. Are quality costs integrated with other quality reports?
9. Does the company have a quality assurance department (coordinator etc.)?
   a. Do the QA personnel have adequate authority?
   b. Do the QA personnel appear to be qualified?
   c. Does the QA personnel's performance appear to be adequate?
10. Does the Company have a written QA program?
    a. Is the QA program suitable for the product or service?
    b. Does the design of the QA program appear sound?
    c. Was a recognized and valid program standard applied?
    d. Is the QA program documented in a Quality Manual?
    e. Was such a standard required by a customer?

QUALITY MANAGEMENT SYSTEM

    f. Was the program audited by an external auditor?
    g. Is the QA program registered by an recognized agency?
    h. Does the company have to comply with several, partially conflicting program standards?
    i. Was the QA program reviewed and audited internally?
    j. Is this QA program audited and is it in compliance with a standard?

DESIGNING THE QUALITY ASSURANCE SYSTEM

1. Was a project or action plan or both prepared and followed?
2. Was the QA department adequately involved?
3. Did other departments, such as Marketing, participate?
4. Did the QA department survey current practices, strengths, and weaknesses?
5. Was an effective project management approach used?
6. Did the QA department have a formal project plan?
7. Were computer-based project planning and control methods used?
8. Were special quality improvement projects carried out?
9. Were these projects well planned and effective?
10. Was the procedure writing properly organized?
11. Were workshops for procedure writing conducted?
12. Were these workshops properly planned and followed up?

PRODUCT/SERVICE DESIGN ASSURANCE

1. Are there QA review procedures for design activities?
2. Are design documents properly planned and controlled?
3. Is the QA department adequately involved?
4. Are other departments informed?

5. Are changes in design documents properly planned and controlled?
    a. Is a system established for monitoring such changes?
    b. Is use of obsolete documents precluded?
    c. Is the status of documents clearly marked?
    d. Is the distribution controlled?
    e. Does QA verify changes?
    f. Is Production/Operations involved?
    g. Is Marketing involved?
    h. Is Purchasing involved?

6. Are contracts with customers reviewed?
    a. Are these reviews performed early enough?
    b. Are adjustments made?
    c. Do contracts have explicit QA stipulations?
    d. Is QA involved in these reviews?

7. Are quality (inspection) plans prepared for each product/service/contract?
    a. Are these plans adequate in format and content?
    b. Are there procedures for preparing these plans?
    c. Do major customers assess and approve these plans?
    d. Are plans assessed with regard to resources and capacities?
    e. Are they augmented by written work instructions?
    f. Are inspection or test methods appropriate?
    g. Are statistical methods used?

SUPPLY ASSURANCE

1. Is a qualified suppliers list kept?

2. Is this list properly maintained?

3. Are suppliers properly selected?

4. Is a supplier's performance evaluated and recorded?

5. Are suppliers informed about QA responsibilities?

6. Are source inspections conducted?

7. Are suppliers' QA programs audited?

8. Is the QA department adequately involved?
9. Are quality and quality assurance requirements properly inserted into the purchasing contract?
    a. Is QA department involved in purchasing decisions?
    b. Are requisitions approved by QA?
    c. Are purchasing contracts checked and approved by QA?
    d. Is QA involved in nonconformance handling?
    e. Are inspection or test reports required by suppliers?
    f. Are these checked and verified by QA?
10. Are receiving inspections performed?
    a. Are there proper procedures for receiving inspection or tests?
    b. Are documents and items checked against contractual requirements?
    c. Is inspection status marked on items and documents?
    d. Are deliveries properly segregated according to status?
    e. Are there procedures for handling nonconformances?
    f. Are inspection and storage facilities adequate?
    g. Is the inspection staff properly trained?

PRODUCTION ASSURANCE

1. Is production readiness verified?
2. Are there procedures for first-piece inspections?
3. Is work-in-progress inspection performed according to plan?
4. Are inspection or test results properly recorded?
5. Does inspection verify proper material and documentation prior to production?
6. Are approved materials identified and segregated from those not approved?
7. Is compliance with work instructions inspected?
8. Are inspection or test plans available?
9. Do these plans specify tests, equipment, test sequence, acceptance criteria, test environment, variable data to be recorded, and so forth?

10. Are special processes controlled in accordance with procedures?
    a. Is subcontracted work properly controlled?
    b. Are operators properly trained and certified?
    c. Are special inspection or tests performed for such processes?
    d. Are there special work and operations instructions?
    e. Do these special instructions include QA instructions?
    f. Are these instructions available?
    g. Is compliance with such instructions supervised?
    h. Are process performance standards established?
    i. Are inspection or test results properly recorded?

11. Are statistical process control methods properly applied?

12. Do operators conduct special QA inspection or tests?
    a. Are operators properly trained and supervised?
    b. Are the inspection or test instructions adequate?
    c. Are these instructions up-to-date and available?
    d. Are the results properly recorded?

13. Is an end-item inspection performed?
    a. Are the inspection or test plans appropriate?
    b. Are the plans complied with?
    c. Are the facilities adequate?
    d. Are results recorded?
    e. Is the inspection status clearly identified?

NONCONFORMANCE CONTROL

1. Are there effective systems and procedures for controlling nonconforming items?

2. Are nonconformances dealt with without delay?

3. Are alerts issued?

4. Is there a recall system or procedure?

5. Are the nonconforming items properly identified, segregated, and disposed of?

6. Is there a special closed-in area for these items?

QUALITY MANAGEMENT SYSTEM

7. Is there a list of authorized and qualified inspectors and decision makers for these items?
8. Is the QA department adequately involved?
9. Is the nonconformance clearly stated on documents?
10. Are causes analyzed and preventive measures taken?
11. Are corrective actions followed up by the QA department?
12. Are those responsible for nonconformances informed about the occurrences and are remedial actions taken?
13. Are there special measures in case of repetitive nonconformances?
14. Are the procedures for repair and rework adequate, available, and complied with?
15. Are scrapped items properly controlled?

QUALITY CONTROL AND INSPECTION RESOURCES

1. Are shop areas and inspection/test facilities conducive to reliable performance and results?
2. Are inspection and test equipment controlled?
   a. Are there special procedures to control this equipment?
   b. Are these procedures adequate and complied with?
   c. Is a register for these items kept?
   d. Are the items identified?
   e. Is a schedule prepared and complied with?
   f. Are valid measurement standards and prescribed procedures applied?
   g. Are subcontractors properly selected and controlled?
   h. Are obsolete items discarded and controlled?
   i. Are calibrations, etc., documented in a log?
   j. Are not-in-use items kept in safe places?
   k. Are items used by subcontractors and suppliers verified?
3. Are inspectors properly qualified and certified?
   a. Is the supervision adequate?
   b. Are performances evaluated and training conducted?

PHYSICAL PRODUCT INTEGRITY

1. Are storage facilities adequate?
2. Are items in storage maintained and checked?
3. Are all items identified?
4. Is there a special storage area for nonconforming items?
5. Are transportation and handling proper?
6. Are items requiring special storage and handling taken care of?
7. Is quality safeguarded during transit?

CUSTOMER RELATIONS

1. Does QA check advertisements and customer manuals and brochures?
2. Is there a complaint-handling procedure?
3. Are customer services adequate?
4. Are instructions for repairs and maintenance adequate?
5. Is the staff supervised?
6. Is QA adequately involved?

QUALITY MANAGEMENT INFORMATION SYSTEM

1. Is the Quality Management Information System documented?
2. Are quality reports prepared?
3. Are records and reports standardized?
4. Do these records and reports plan and control data compilation?
5. Are data analyzed?
6. Is information available?
7. Is information retrievable?

8. Is the reporting scheme periodically reviewed?
9. Are confidential matters safeguarded?
10. Are records retained?
11. Does the system allow tracing of events and decisions?
12. Are computer facilities utilized?

STATISTICAL QUALITY PLANNING AND CONTROL

1. Are statistical methods applied?
2. Do training programs include statistical methods?
3. Do QA specialists have adequate knowledge of these methods?
4. Are design of experiments or the Taguchi method or both applied in design activities?
5. Are reliability analyses performed?
6. Is statistical process control applied?
    a. Are process capability studies performed?
    b. Are supervisors and operators trained in statistical process control?
    c. Are control charts properly applied?
    d. Are technical control chart standards used?
    e. Are control charts used discriminately?
    f. Are the proper kind of control charts selected?
    g. Are advanced types of control charts also used?
    h. Is the application of control charts approved by QA?
    i. Is the sampling frequency adequate?
    j. Are the data validity and reliability ensured?
    k. Are information from control charts monitored, communicated, and utilized for QA?
    l. Are records retained and traceable?
7. Is acceptance sampling applied?
    a. Are acceptance sampling standards, e.g., MIL 105D, used?
    b. Is the application valid and reliable?
    c. Are the parameters, e.g., acceptable quality level, reasonable?

d. Is the method used discriminately?
e. Is the application verified and approved by QA?
f. Is training provided when the method is applied?
g. Is the method properly inserted into the inspection or test procedure?
h. Are advanced types of acceptance sampling also used?
i. Is the average outgoing quality level method applied?
j. Do observations and results lead to a review of the sampling plans?
k. Is the application by suppliers properly verified?
l. Are records retained and traceable?
m. Is the application of acceptance sampling reviewed periodically?

SOFTWARE QUALITY CONTROL

1. Is there a software quality control program?

2. Does this program comply with applicable standards?

3. Is there quality assurance verification of procured software?

4. Do they have a system or procedure for requisitioning, selecting, and introducing software?

5. Are there formal review procedures for the following:
    a. Software requirement review?
    b. Preliminary design review?
    c. Critical design review?
    d. Software verification review?
    e. Software integration review?
    f. Application review?

6. Are software packages properly stored, easily retrievable, and properly maintained?

7. Is there a procedure for implementing software changes?

8. Is obsolete software eliminated and its disposal controlled?

9. Is there a disaster recovery procedure?

10. Is there a procedure for hardware maintenance?

QUALITY MANAGEMENT SYSTEM

11. Is there a special QA department for software QA?
12. Are the personnel qualified and certified?
13. Is training provided to ensure proper application?
14. Is documentation properly planned and controlled?

IMPLEMENTING THE SYSTEM

1. Are there adequate procedures for implementing the QA procedures?
2. Is the implementation reviewed or audited?
3. Is there a procedure-change procedure?
4. Are sufficient documentation and support provided for implementation, e.g., Quality Manual, workshops?

QUALITY MANUAL

1. Is the manual approved and signed by senior management?
2. Does the manual comply with any program standard?
3. Is there a signed policy statement?
4. Does the manual have a periodic review and change page?
5. Is there a manual review and change procedure?
6. Does it have a distribution list?
7. Does the manual include:
   a. QA organization description?
   b. Procedure for test or inspection plans?
   c. Documented procedures for
      i. Design review?
      ii. Contract review?
      iii. Document control?
      iv. Measuring and testing equipment?
      v. Supply assurance?
      vi. Production assurance?
      vii. Identification and traceability of items?

       viii. Special processes?
        ix. Handling, storing, and shipping?
         x. Nonconformance control?
        xi. Quality records?
       xii. Audits?

8. Is the format of the Manual adequate?
   a. Title page?
   b. Page layout?
   c. Appendix?
   d. Cross references?

WORKSHOPS

1. Are workshops properly planned?

2. Is a workshop plan prepared, approved, and communicated?

3. Are workshops integrated with other training arrangements?

4. Are instructors properly selected and supervised?

5. Are objectives and methods established?

6. Are learning materials adequate?

7. Are tests adequate?

8. Are results certified and recognized?

9. Are workshops evaluated by participants?

10. Are workshops followed-up with the participants?

11. Are records on workshop content, etc., kept?

12. Are training needs assessed periodically?

13. Are workshops reviewed periodically?

AUDITING QUALITY ASSURANCE

1. Do audits appear well planned and effective?

2. Are there a formal audit policy and a program?

3. Are all areas of QA periodically audited?

QUALITY MANAGEMENT SYSTEM

4. Are audits conducted for special reasons?

5. Do the audits apply and comply with audit standards?

6. Are computer facilities used for auditing?

7. Are audit standards and guidelines available to auditors?

8. Is the audit function well organized?
    a. Are auditors trained and certified?
    b. Are continuous training and the updating of the qualification adequate?
    c. Are auditors independent?
    d. Is their performance evaluated?
    e. Do they have access to experts?
    f. Are audit teams well organized?
    g. Is the status of the audit function adequate?
    h. Are human and financial resources adequate?
    i. Do auditors participate in the auditor associations and committees?

9. Does management initiate and support audits?

10. Does management receive reports and act upon them?

11. Are audit objectives formalized?

12. Are both the system effectiveness and the compliance with standards examined?

13. Are audit tasks properly assigned to auditor(s)?

14. Are audit teams properly organized?

15. Are internal audits adequate?

16. Are external audits, e.g., supplier, adequate?

17. Are audit plans prepared and communicated?

18. Are checklists and other working papers adequate?

19. Are preaudits performed, e.g., check of quality manual?

20. Are schedules and meetings approved by auditee?

21. Is the audit announced properly?

22. Are audit methods adequate, e.g., is statistical sampling used?
23. Are checklists properly used during audits?
24. Are observations:
    a. properly recorded?
    b. facts on observations compiled?
    c. properly communicated?
    d. properly assessed by lead auditor and reported?
    e. followed up?
25. Are preaudit and postaudit meetings held with auditors?
26. Are records of these meetings prepared, signed, and retained?
27. Are there procedures for involving experts in the audit?
28. Are confidential matters safeguarded?
29. Does the audit style encourage auditees to participate and support the audit?
30. Do audit reports
    a. have the proper format?
    b. have the proper content?
    c. receive proper dissemination?
    d. have their intended impact?
    e. have the proper follow-up?
    f. contain additional reports on corrective action?
31. Are audit documents properly retained and accessible?
32. Is audit performance evaluated and communicated?
33. Is there a procedure for quality assurance of audits?
34. Are audits integrated with audits
    a. by customers?
    b. by regulatory bodies?
    c. by registration agencies?
    d. by corporate auditing branch?
    e. with financial audits?
    f. with internal audits?

6
CASES

Three cases have been selected to demonstrate different situations. In the first case a small hospital wants to establish a quality assurance program. This is a fairly unconventional setting. Many new approaches have to be devised in which audits play an important role.

In the second case a manufacturing firm has a quality assurance system that is ineffective and major improvements have to be made. Often is it more difficult to change an outdated system than it is to design and implement an entirely new system. Here, again, planning proceeds carefully utilizing input from frequent audits.

A research establishment is the third case. All phases of a research project have to be planned and controlled with regard to the quality of the work performed and the results achieved. This unusual but realistic situation demands a creative approach.

6.1 Hospital Quality Assurance System

INTRODUCTION

A 200-bed community general hospital in a small northern town in Canada applied for accreditation by the Canadian Council on Hospital Accreditation (CCHA). The CCHA standard requires a quality assurance system. The hospital board set the following objectives, leading toward accreditation:

- ★ Quality assurance (QA) coordinator is appointed and reports to a newly formed executive committee on quality assurance.

- The quality assurance coordinator develops a quality assurance system in accordance with applicable standards and in cooperation with supervisors and medical staff.
- Audits accompany the system development, to be arranged by the QA coordinator.

The coordinator decided on the following approach:

1. Determination of system and audit principles and definitions.
2. Compilation, study, and selection of applicable standards and development aids.
3. Clarification of audits in support of planning and system design.
4. Design of the system.

SYSTEM PRINCIPLES

The principles adopted for the planning and auditing of the QA system are consistent with the standards imposed by the accreditation body and complementary ones selected by the QA coordinator. The main principles are:

- project format for stepwise system development,
- supervisors for their respective area establish and chair a project team,
- staff participates in the project, and QA coordinator acts as a facilitator,
- standards and other aids are applied with the assistance of the QA coordinator,
- project is initiated with a baseline audit, and subsequent audits control project progress,
- outcome of projects are quality assurance procedures,
- QA coordinator prepares a Quality Manual.

DEFINITIONS

Definitions applicable to the quality of health services are given in the following paragraphs.

Quality means the greatest achievable health benefits, with minimum unnecessary risk, and use of resources in a manner satisfactory to the patient (1975 Forward Plan for Fiscal Years 1977–1981, U.S. Department of Health).

Quality of care consistently contributes to improvement and maintenance of the quality or duration or both of life.

Quality assurance is the comprehensive planning and control of product and service quality. Companies and, more recently, health care facilities design and implement their quality assurance organization, principles, and procedures as a coordinated, more or less comprehensive program. Standards, guides, and other technical support provide important incentives and assistance. The following are relevant definitions for the health care service:

* *Quality assurance* is the establishment of hospitalwide goals, the assessment of procedures in place to see if they achieve these goals, and, if not, the proposal of solutions to attain these goals. The quality assurance program should be internal, internally administered, ongoing, specific to the institution, structured, and coordinated within the facility. An audit is considered a part of the QA program, as defined here.

* *Quality assurance* is a process in which standards describing the level of quality desired and feasible are set, the level of achievement of those standards is measured, and action is taken to correct identified differences, according to the College of Nurses of Ontario. Here, again, audits are an integral and essential element of the quality assurance program.

Audit in the context of this chapter and in accordance with applicable generic auditing standards is a documented activity performed to verify, by examination and evaluations of evidence, that applicable elements of the quality system are appropriate and have been developed, documented, and effectively implemented in accordance and in conjunction with specified requirements.

The quality system embraces all acts to ensure quality. Audits require well-defined performance standards and the existence of a formal quality assurance program.

Audit in the health care service is not as clearly defined as it is necessary and desirable for effective auditing. An audit in a health care setting is a type of patient medical care evaluation study in which the indicators of performance are compared with performance and outcome documented in patient medical records. Records are reviewed by a committee of peers, and if deficiencies are noted, corrective actions are instituted. The audit should produce results that reflect the quality of care provided to a significant proportion of the hospital patients with a specific admission or discharge diagnosis. To avoid confusion, one should reserve the term *audit* to mean a specific study method for reviewing patient records against clinical criteria.

Audits in the health care industry seem to mean various kinds of evaluations from retrospective case studies, over current "chart" controls, to comprehensive reviews and studies mainly in connection with accreditation formalities or research projects. The resulting confusion in semantics and practice of evaluations calls for some basic clarification, such as generic standards have to offer.

STANDARDS FOR SYSTEM AND AUDITS

The QA coordinator selected the Quality Assurance Standard of the Canadian Council of Hospital Accreditation, in which one substandard deals explicitly with quality assurance. Standard VIII Quality Assurance of the Canadian Council of Hospital Accreditation states: The health care facility shall demonstrate a consistent endeavor to deliver optimal patient care. A major component in the application of this principle is the operation of a quality assurance program.

This Standard permits multiple approaches to detect and assess problems and to monitor the effectiveness of changes. The quality assurance system must be documented, coordinated by appropriate personnel, and operated with wide staff participation. The system should be audited annually.

Because the CCHA Standard leaves much open to interpretation, the QA coordinator adopted the *Quality Management Guidelines* (ANSI/ASQC, Q94-1987) and the *Generic Guidelines for Auditing*

Quality Systems (ANSI/ASQC Q1-1986), American Society for Quality Control, Milwaukee, WI, as additional aids.

OTHER INFORMATION SOURCES

Considerable uneasiness was felt with regard to continuous auditing. Performance evaluations, studies, and peer reviews had been introduced with little success, and the audit seemed to run in the same direction. A proliferation of such assessments, even when used for planning and maintaining the QA System, would be resisted mainly by the professional staff members.

The QA coordinator conducted a special study on auditing in health care facilities and compiled the following bibliography:

1. Millenson, Michael L., A Prescription for Change, *Quality Progress,* May 1987, p.16f.

2. American Medical Association, Council on Medical Service, Quality of Care, *Quality Progress,* May 1987, p.22f.

3. Ryan, John, Health Care Quality Assurance Regulation, *Quality Progress,* May 1987, p.27f.

4. Millenson, Michael L., Making Health Care Measure Up, *Quality Progress,* May 1987, p.35.

5. Balfe Bruce E., et al., A Health Policy Agenda for the American People, *Quality Progress,* p. 48f.

6. Haffner, Alden N., Jonal, Steven, Pollack, Burton, *Regulating the Quality of Patient Care,* in Pena, J. J., Rosen B., Haffner, A. N., and Light, D. W., *Hospital Quality Assurance,* Aspen Systems Corporation, Rockville, Maryland, 1984.

7. Willborn, Walter, Quality Assurance Audits and Hotel Management, *The Service Industries Journal,* Nov. 1986, p.293f.

8. Pena, Jesus, Haffner, Alden, Rosen, Bernard, Light, Donald, *Hospital Quality Assurance,* Aspen Systems Corp., Rockville, Maryland, 1984.

9. Donabedian, A., Evaluation the Quality of Medical Care, *Milbank Memorial Fund Quarterly,* 44, part 2, 1966, p.166f.

10. Greenspan, Jack, *Accountability and Quality Assurance in Health Care,* Charles Press Publishers, Bowie, Maryland, 1980.

11. Canadian Council On Hospital Accreditation, *Proceedings of the Seminars on Quality Assurance,* Ottawa, Ontario, October 1983, May 1984.

12. Reenen van, J. A., Quality Assurance: introduction to terminology and literature, *Hospital Trustee,* Nov./Dec. 1983, p.18f.

13. Sanazaro, P. J., Quality assessment and quality assurance in medical care, *Annual Review Public Health,* No. 1, 1980, p.36.

14. Marshik-Gustafson, J., et. al., Planning is the key to successful QA programs, *Hospitals,* Noo. 11, 1981, p.67.

15. American Society for Quality Control, Quality Auditing Technical Committee, *American National Standard, Generic Guidelines for Auditing of Quality Systems, ANSI/ASQC Standard Q1-1986,* American Society for Quality Control, 310 West Wisconsin Ave., Milwaukee, Wisconsin.

16. Sinha, M., Willborn, W., *The Management of Quality Assurance,* John Wiley & Sons, Inc., New York, 1985.

17. Emergency Health and Ambulance Service Section, Manitoba Health Services Commission, *Policy and Procedure Manual, Guideline,* Manitoba Health Services Commission, 1983.

18. JCAH, *Quality Review Bulletin (monthly),* Joint Commission on Accreditation of Hospitals, 875 North Michigan Ave., Chicago, Illinois.

19. ASQC, *Standards,* American Society for Quality Control, 310 West Wisconsin Avenue, Milwaukee, Wisconsin.

20. Willborn, Walter, A Generic QA Audit Guideline, *Quality Progress,* January 1987, p.24/25.

21. Willborn, Walter, *Compendium of Audit Standards,* American Society for Quality Control, Milwaukee, Wisconsin, 1983.

22. Canadian Association of Quality Assurance Professionals, *QA: Beyond Theory into Practice,* Eighth Annual Conference, Toronto, Ont. October 14–16, 1987, CAQAP, Suite 480, 151 Bloor Street West, Toronto, Ontario, Canada, M5S 1T3.

The executive committee agreed to apply generic system and audit standard where feasible. The QA coordinator in using these standards in addition to and consistent with the mandatory standard of the accreditation body broke much new ground. Because of the principles adopted for the system design, in particular the decisive participation of staff, the various project teams functioned harmoniously and constructively.

AUDIT PRINCIPLES

Quality assurance audit, or *quality audit* is a formal and systematic evaluation of the Quality Assurance System, especially compliance with applicable standards and the System's effective design and implementation. The object of the audit is predetermined and can be the entire quality assurance system, or individual procedures, product-based systems, contracts, or processes in the hospital.

Auditing of quality assurance and quality of services has important benefits for the health service facility such as:

* Auditing is welcomed as a defense against evaluations on price and cost alone.

* Auditing justifies cost and fee increases.

* Auditing facilitates performance-oriented payment schemes without compromising quality.

* Auditing restricts unnecessary services.

* Auditing creates trust and confidence in the staff and public community.

* Auditing provides protection against malpractice suits.

* Auditing assists in peer reviews and their resulting decisions.

In view of the various quasiaudits the QA coordinator listed different types of audits for better clarification. Basically each audit is unique with regard to objectives, setting, plan, staff involved, results, and final outcomes. Certain types of audits have a greater impact on quality of health services than others and consequently demand more planning, care, auditor competence, time, and funds.

Relevant audit types were baseline, internal and external, and special audits. A baseline audit established grounds for subsequent special and regular follow-up audits. It is crucial in determining important prerequisites for future auditing and helps to plan an effective quality assurance program. Internal and external audits are of the regular type, but external audits are usually important for accreditation of the facility. Both internal and external audits complement each other and require close cooperation of auditors. Finally, special audits assist research and similar study projects, in order to ensure the validity, reliability, and fairness of the outcome. Again, special audits complement such similar evaluation and research projects.

Three kinds of audit objects were distinguished:

* structure: facilities, equipment, etc.;

* processes: housekeeping, nursing, surgery, etc.;

* outcome: impact on patient health and health care.

Important prerequisites for effective auditing were that quality assurance procedures and the resulting documentation and reports were available, quantitative quality measures had been defined, and auditor and auditee were well prepared and agreed on audit standards, objectives, and principles.

The essential auditing methods applied were sampling from evidential material, checklisting, interviewing, and observing. Reports were widely distributed with follow-up measures decided by the respective supervisor. A copy of the report was sent to the executive committee.

As the auditor, usually the QA coordinator or a supervisor from another area, evaluated compliance and studied possible procedural improvements, the following were the leading questions in each audit element:

1. Does a procedure exists?

2. If yes, was the procedure invoked or applied?

3. If yes, was the performance effective?

4. If yes, confirm the procedure.

5. If the answer(s) is (are) no to any of the above questions, the performance and actual practice are assessed for a positive or negative outcome and impact on quality and quality assurance.

6. If the practice is accepted, a procedure is either established or modified. The QA coordinator adopted the "Basic Audit Procedure" explained in this Guide.

SYSTEM DEVELOPMENT

In each department separate quality assurance procedures were formulated by the project teams. After the baseline audit, individual problems were defined, discussed, and assigned to a group for solving. The QA coordinator ranked these tasks, provided technical support, and guided progress by means of frequent miniaudits. Much effort was spent in workshops—conducted like Quality Circles—and procedure writing. Once the departmental QA procedures were established, their effective implementation was audited. After any additional modification, the final procedure was submitted for approval by the executive committee and incorporated into the official Quality Manual of the institution.

When the Quality Manual was complete—all procedures and required system elements successfully implemented—the accreditation body was informed and an external audit was requested. The Quality Manual explained the new quality assurance system to these auditors, which assisted proper audit planning. An outline of the Quality Manual is included in the Appendix of this Guide.

The QA coordinator selected the housekeeping department for the pilot project and the writing the first QA procedure. The baseline audit observed that the laundry was not properly cleaned, rooms and beds were not ready in time when new patients arrived, safety and hygienic standards were not adequately complied with, and the staff lacked proper work instructions. There were some other, minor observations that the staff had already corrected.

The supervisor held a meeting with the department's staff and the QA coordinator, who explained the applicable quality assurance standards and principles. A project team then was formed for drafting a quality assurance procedure that embodied current good housekeeping practices and ascertained the necessary improvements required. The following is the drafted Procedure.

EXAMPLE FOR A PROCEDURE: QUALITY ASSURANCE IN HOUSEKEEPING

PURPOSE

The department and staff will be guided by this standard procedure in their efforts to maintain good housekeeping. This standard contributes to the quality assurance; safety of patients, visitors, and staff; and the general quality image of the hospital.

POLICY

Quality assurance of the various services in and by the hospital embraces all departments, functions, and jobs. In housekeeping various suitable and approved planning and control methods are applied for consistently achieving and maintaining good housekeeping practices. All staff participates in this effort with supervisors and QA coordinator providing support. A procedure explains the guidelines and principles. Audits verify compliance and effectiveness.

APPLICATION

This standard applies to supervisor(s), all housekeeping staff employed by the department, and the QA coordinator.

DEFINITIONS

Good housekeeping comprises the decisions and activities resulting consistently and effectively in clean, safe, and hygenic physical environment for patients, visitors, and hospital staff and that meets applicable standards.

Quality assurance of housekeeping includes the procedures, practices, recommendations, and other efforts for improvement that are directed toward "good housekeeping." Housekeeping does not include food services and nursing care.

REFERENCES

The applicable references include Quality Assurance Standard of the Canadian Council on Hospital Accreditation, and applicable technical standards.

STEPS/METHOD

The following measures are taken by the supervisor in cooperation with staff and QA coordinator.

1. Audit current practices and the compliance to instructions. Baseline audit results are followed-up by supervisor and staff with support of the QA coordinator.

2. Prepare work instructions and assignments, job descriptions outlining qualifications and responsibilities, and explicit quality assurance directives.

3. Discuss and clarify written documentation on individual and collective quality assurance, and finalize this documentation. Check with applicable standards and obtain approval.

4. Assign tasks for implementing and further improving quality assurance of housekeeping. Provide support and conduct meetings. Allow for modifications.

5. Prepare checklists and statistical control charts for performance control and audits.

6. Conduct audits in accordance with audit standards, emphasizing compliance and system improvements. Audit in regular intervals or when needed.

7. Review quality assurance documentation and staff qualification and performance, based on audit results.

These steps can be altered, of course, to reflect different conditions.

FORMS

Applicable forms for this procedure include work instructions and job descriptions, incidence ("trouble") reports, checklists, and statistical quality control charts.

EXPLANATORY NOTE

The major principles of quality assurance in housekeeping are:

* explicit quality assurance directives,
* participation of staff in planning and control,
* self-inspection and statistical quality control wherever feasible,
* emphasis on preventing defective housekeeping,
* QA coordinator provides technical assistance,
* audits enhance supervision and system effectiveness.

The major benefits from compliance to this Procedure and continuous effort to maintain and improve quality assurance are:

* enriched jobs,
* recognition of good performance,
* contribution to quality assurance effort of the hospital and sharing of the benefits,
* raising the status of housekeeping with subsequent better cooperation from staff,
* proper orientation and introduction for new staff.

This Procedure served as a documented framework for the detailed work planning and control. Formal approval by the Executive Committee and the various auditors increased staff acceptance and compliance. The QA coordinator incorporated this Procedure into the Quality Manual and used this as a practical example for other departments to follow. (See partial draft of the Quality Manual with other Procedures still to be prepared in the Appendix of this Guide.)

Self-inspection was successfully introduced in the laundry along with statistical control charts for defects ("c-chart"), which were set up by the QA coordinator. For the job of room cleaning for a new patient, a checklist was prepared for the operator, supervisor, and auditor. This book's Procedure for self-inspection was applied.

Cost containment, a major issue in most hospitals, did not affect the quality assurance effort. On the contrary, the QA coordinator prepared a quality cost study along with a regular quality report that convinced management about the rationality of the new quality assurance system. The hospital accreditation audit has not been requested since the procedures for other departments must still be prepared.

The combined planning and audit approach proved useful mainly because audits revealed strengths and weaknesses at crucial points in the system development. The staff could communicate, sometimes confidentially, their feelings about the changes and make recommendations.

6.2 A Small Manufacturer's Quality Assurance System

INTRODUCTION

The owner/manager-operated plant employs 110 persons, most being skilled craftspersons. The firm produces large electric motors for industrial customers and for electrical power generating utilities. It has been located in Milwaukee, WI, since 1945.

A chief inspector—a qualified quality engineer—prepares all inspection and test plans and supervises three inspectors. A Quality Manual describes the inspection system that complies with a military standard, MIL-I-45208. The production manager has frequently overruled rejections by inspection in order to meet deadlines. The chief inspector is quite frustrated. Customer complaints about poor quality have been sent directly to the owner/manager.

A major utility conducted an audit of the inspection system. Not surprising to the chief inspector major problems were noted, and cancellation of contracts has been threatened if corrective actions were delayed.

The owner/manager reacted immediately by appointing the chief inspector as quality assurance manager with the task of developing a satisfactory quality assurance program. The production manager remains fairly skeptical, but agrees that something has to be done.

PREPARATION FOR SYSTEM DESIGN

Some strategic measures had to be taken before the system design could proceed. The position of the quality assurance manager, former chief inspector, had been strengthened and full support of the owner/manager was forthcoming. It was decided in a meeting attended by all managers and supervisors that the current inspection system had to be improved so that the next external customer audit would be passed. This audit was to occur within three months. The quality assurance manager chaired the quality assurance review and audit committee.

As demanded by the customer, the new quality assurance system was to comply with the Mil-Q-9858A, *Quality Program Requirements*. Generic guidelines were also used, such as the ANSI/ASQC Q94-87 or ISO 9004 documents, mentioned frequently in this book.

The owner/manager wanted the quality assurance manager to arrange frequent audits in order to be kept well informed about the progress being made. The *Generic Guidelines for Auditing of Quality Systems* (ANSI/ASQC Q1-1986) was approved for this purpose.

The quality assurance manager welcomed the auditing mainly for the following reasons:

* proper follow-up of previous external audit observations and the resulting corrective actions;
* initiating defect prevention rather than mere correction of detected weaknesses;
* using own audit observations in workshops and for assigning tasks to project teams;
* opening communication channels with individual operators;
* overcoming departmental boundaries and providing supervisors with a broader outlook by involving them as auditors for areas other than their own;
* evaluating actual project progress and redirecting the effort without delay;

* preparing checklist and other working papers for the audit and auditee that also allow for preparatory self-audits (checks);

* preparation for the next external audit(s);

* and better continuous and more official communication with the owner/manager.

GENERAL OUTLINE OF DESIGNING THE SYSTEM

The quality assurance manager started with a fact-finding audit; actually a follow-up audit of the preceding disastrous customer audit. This helped to clarify the observations made and to establish more concerted and informed effort for correction and prevention. Moreover, the objective of establishing a quality assurance system required a more comprehensive approach than just correcting detected weaknesses. The former, mainly reactive approach had to be replaced by an approach with considerable staff input. The new System Standard was imposed and strict compliance with it was required.

The major observations of the preceding audit had been:

* subcontractors are not being properly informed, supervised, and audited;

* end-item inspection results had been overruled by the production manager, resulting in unnecessary returns;

* nonconformance handling was ineffective with regard to defect prevention and no restricted area was set aside;

* inspectors used test equipment not properly calibrated and controlled.

Three projects were formulated and given a priority ranking:

1. Clearance of all external and internal audit observations.

2. Development of additional system elements as required by the new standard.

3. Preparation for the next external audit and cooperation with the auditors.

Audits accompanied all three projects at predetermined milestones. Project management methods were also applied for the audits. This meant setting of clear objectives in terms of expected outcome and timing, suitable organization and representative membership, meetings with thorough planning and performance controls, and flexibility to change in order to achieve the project goal optimally.

After the projects were underway, the quality assurance manager conducted additional workshops on quality assurance principles and methods with direct reference to the problems that had been encountered and the questions that had arisen. Quality circles offered a more congenial forum for interpersonal exchanges. A member of the new quality assurance department attended all these meeting, which were chaired by the supervisor.

New quality assurance procedures were prepared with reference to the Standard and this Guide. In each case the quality manager involved the individuals concerned and used the procedure writing for further instruction on quality assurance matters. Once procedures reflected proper practices and were understood and accepted by supervisors and operators, they were officially incorporated in the new Quality Manual. (A partial outline of this Manual is given in the Appendix.)

Because of the limited time available, three months, all procedures were considered preliminary subject to further modifications.

Shortly before the deadline, the quality assurance manager conducted an audit using the preliminary procedures. Checklists had been prepared and distributed. Any problem had to be resolved immediately because of the pending external audit.

One conflict between the production manager and the quality assurance manager happened during the final phase of system development. A motor had been rejected by inspection, because vibrations exceeded the upper control limits of the newly instituted statistical quality control chart. The production manager wanted to reinspect it, because it was felt that the customer would not notice the vibration. The quality assurance manager, to whom the inspector reported, refused to sign the clearance papers, without which the delivery could not be made.

The production manager complained about this unnecessary bureaucratic interference by inspection to the owner/manager without avail. There had been agreement that all new procedures were to be complied with and audits were to be used to make changes. There was no justification in this case to circumvent the procedures, even when they were declared "preliminary." The quality assurance manager was glad about the owner/manager's support, and the production manager accepted the decision reluctantly. The message spread very quickly in the plant and among the customers.

Some customers, not requiring compliance with the system standard, considered the new program extraordinarily costly and unnecessary. As there was no other supplier readily available, they feared that prices would rise.

The marketing manager pointed out that quality cost report showed a considerable reduction of failure and total costs. The MIL-9858A Standard stipulates quality cost data as a management element. Higher productivity due to less failures would allow a price reduction once the system development costs were recaptured.

The deadline was met with one exception: major suppliers had to be given more time to adopt this standard, so that the quality assurance manager did not have enough time to conduct all the supplier audits that were required. The new supplier relations and auditing procedure, nevertheless, was accepted by all and agreement was confirmed.

The external lead auditor realized the predicament of the quality assurance manager. The audit confirmed, however, that the system development was successful to that point and the validity of the Quality Manual. The visible commitment of the owner/manager and the senior staff, the competence of the quality assurance manager to use audits as a management method, and the involvement of the operators in quality assurance did not go unnoticed by the external auditors. A follow-up audit was scheduled. Internal and external auditors worked closely together. Sound internal auditing in combination with quality assurance planning allowed external auditors gradually to reduce their auditing frequency.

6.3 Quality Assurance in a Research Establishment

INTRODUCTION

Quality assurance of design activities and of computer software has been introduced recently as the need for it was perceived.

Ensuring the quality of services has become another area for the application of concepts and methods that had been used mainly in manufacturing. The importance and impact of basic and applied research demands that the quality of major research projects is to be planned and controlled.

Well-proven quality assurance procedures from manufacturing and elsewhere cannot and should not directly be transferred and applied to research activities. As the term "research" indicates, a researcher searches for new knowledge and for explanations and solutions to problems of all kinds.

One can still distinguish, however, sound and good research from that which is less impressive and less convincing. The quality of research output depends to a certain extent on the qualification of the researcher(s), the resources available, the organization and its management, and the general support and acceptance of the efforts and projects. Quality assurance therefore seems to have a role to play in the research field as it does in other realms.

The following is to guide the design and implementation of a quality assurance program for research projects. This must be carefully performed in order to promote "quality research." New principles need to be formulated so that researchers understand and appreciate the quality assurance procedures and program.

There is not much literature on this special topic. In the explanatory note we report on some of the publications. This Guide has also assisted managers in this unfamiliar area of quality assurance application.

PURPOSE

Results of R&D projects need to be valid and reliable. Individual researchers faced with complex tasks often want and need reassurance in their work. Evaluators and users of R&D require and demand that certain quality controls during research have been instituted and followed. Quality assurance of research and development projects serves researcher, sponsor, and client to attain maximum satisfaction. It is to support researcher's creativity, to minimize avoidable errors, to provide additional knowledge and experiences, to encourage financing, and to spread the risk of unsuccessful outcomes. It is not to stifle research through unnecessary and unjustified paperwork.

POLICY

A R&D project of a certain scope, complexity, and significance must have an explicit subsystem (procedure) for quality assurance. This procedure is to enhance the value of the research activities. Quality assurance starts with the conception of the project and accompanies research activities through the planning, execution, review, and application stages. Researchers with the assistance of quality assurance specialists and established guides should carry out planned and approved quality controls.

ORGANIZATION

A QA department is required for designing, implementing, and maintaining the QA program. Organizational charts and written descriptions differentiate authority and duties. Sufficient freedom for handling problems in research projects must be given. QA management must be independent from the research operations and should report regularly to senior management. Proper status of the QA department enhances the effectiveness of the quality assurance.

QA PROGRAM

In addition to the technical project plan, a special QA plan is to be prepared and approved prior to the start of research. These plans reflect the current knowledge, experience, and expectations about the project and need frequent review and updating. Implementation and documentation are included.

The overall QA program stipulates design controls, procurement, procedures and instructions, material and equipment, traceability, special processes, inspections and tests, nonconformance handling, corrective actions, records and communication, and audits.

APPLICATION

Quality assurance principles and procedures apply to all designated R&D projects. Individual researchers might use proven and available quality control techniques and aids in their work.

Sponsors and clients of the researcher or research team should utilize available and recognized quality assurance aids. Joint planning and controlling of their application best ensures success.

DEFINITIONS

Quality is the totality of features and characteristics of a R&D project that bear on its outcome and the satisfaction with the outcome. Quality is determined by the soundness of concepts, plans, resources, and approaches and methods. Quality is not identical with or identified as expected outcomes.

Quality assurance is all those planned and systematic actions necessary to be confident that a R&D project will satisfy given requirements for quality. Quality assurance stipulations and procedures augment and support quality controls that are inherent to any sound research. Quality assurance in manufacturing is not directly transferrable to R&D, but its essential elements are applicable. The term "quality assurance" is distasteful to many researchers and could be replaced by "quality improvement" (Ref. 4).

Quality plan, quality audit, and quality system are related terms.

R&D project is all interrelated activities that are directed to a common goal. A project is usually performed by a team with a set organization and administration. Project phases are conceptionalizing, planning, executing, reviewing, and reporting. R&D projects have unique risks and uncertainties and demand particular care.

METHODS/APPROACHES

The following steps reflect the normal approach in research activities, but in each of the following steps special quality and quality assurance aspects, procedures, and instructions will augment directives and guidelines.

1. Determine the research goal.
2. Analyze the risks and the resources required.
3. Formulate research project principles, organization, and requirements.
4. Organize the research team.
5. Budget the project.

6. Establish the Project Plan, including the Quality Assurance Plan, using PERT/CPM computer software.

7. Establish the computerized information system, data collection, and reporting schemes.

8. Establish the quality assurance milestones and their respective methods and procedures.

9. Approve the project.

10. Implement the project.

11. Monitor progress; resolve problems.

12. Audit the project at predetermined intervals or milestones.

13. Report the research outcome.

14. Follow up on the research outcome.

STANDARDS

Generic and technical quality assurance program standards can be applied with proper care and modification. Special quality assurance standards for research activities and projects do not yet exist. The Canadian Standards Association currently prepares a standard for projects and project management. A consensus standard for QA of research is needed and should be developed, and the Department of Energy would benefit from this according to Ref. 4.

Roberts (Ref. 2) provides guidelines "for obtaining research and development results of a consistent and known quality." Much of the text tailors standard QA systems, using levels or project QA plans.

FORMS AND TECHNICAL AIDS

Roberts (Ref. 2) has published many examples for research centers, for example:

* Quality Assurance Project Assessment, form and worksheet,
* R&D Project Plan and Status Report,
* Laboratory Notebook,

* Quality Assurance Plan and detailed index of content,
* Purchase Requisition and Receiving Reports,
* Inspection Checklists,
* Corrective Action Report,
* Instrument Service and Calibration Log,
* Technical Procedure (for Tests),
* Independent Review of Project,
* Project Progress Reports,
* Audit Checklist.

Interaction of QA with project activities are shown in a flow diagram starting and ending with the project owner. See example in Ref. 2, p. 79.

GENERAL QA MANAGEMENT OF RESEARCH PROJECT

The U.S. Department of Energy (DOE) has determined that QA principles can and should be applied to all the work it sponsors. It arrived at this position on the evidence of successful use of quality assurance by industry (Ref. 4).

PRECONDITIONS

Three preconditions for quality assurance of research projects are (Ref. 1)

* a concretely formulated goal,
* a mandatory budget,
* a fix deadline for completion.

Research projects must be planned and structured differently than those for development, production, construction, etc., because the typical scientist's behavior is unique.

PROJECT MANAGEMENT SCHEMES

Computerized networks with interrelated work packages and milestones should be prepared. This approach also allows for a quick reaction to unforseen developments in the experimentation, otherwise project management and quality assurance would stifle flexible and creative rethinking and adaptations. The project management scheme and the quality assurance subsystem help scientists to remain goal oriented in their work. Schimmel and Schramm (Ref. 1) demonstrate a general project structure used in the Battelle organization.

OBTAINING THE COOPERATION OF SCIENTISTS

"When an attempt is made to apply proven quality methods in new contexts, the efforts often meet indifference, noncooperation, or even hostility—and mutual frustration" (Ref. 4). The understandable skepticism of researchers in matters of quality assurance of their research work must be overcome by those introducing the quality assurance program.

A scientist strives for originality and creativity. A fixed project management scheme is often seen as an unwanted intrusion or imposition. There is some justification for this attitude. QA is to support the scientist and the project scheme. One serves this purpose best when QA is adapted to the scientist, rather than the scientist being forced into a rigid program. Convincing instead of pursuading or even dictating leads the scientist to acceptance and cooperation.

Directives and compliance to these remain a necessary element of research projects. Scientists will accept these under certain conditions. This is true even for independent external reviews at certain milestones or predetermined deadlines.

QA INTEGRATION

Quality assurance extends to development and production research and activities. Such integration influences and strengthens quality assurance in individual phases.

In a hierarchy of projects we distinguish between general and specific quality assurance measures. Policies provide general guidance and direction; the specific items provide technical procedures and work instructions.

QA MANUAL

A Quality Assurance Plan describes project-specific quality assurance measures and activities. Procedures for purchasing, contracting, inventory, hiring, equipment control and maintenance, project management and administration are more or less valid for all research projects and therefore the Quality Assurance Manual describes these.

Although a research project is similar to any other project, generic standards for quality assurance systems do not apply fully. For instance, the quantities of supplies are relatively small, the technology is complex and new, and the staff is inexperienced in the matter under research. In research there are only few controlled processes, if any. Consequently, available standards, guides, aids, and methods need to be adopted carefully and often need to be modified.

INFORMATION SYSTEM

The project management scheme is an important infrastructure for decision making when relatively high risks and uncertainties are present. All concepts and methods of modern, computer-based project management apply to research projects and to the quality assurance subsystem. In the Battelle organization the standard project scheme and a manual document the expected general research process, including associated quality assurance (Ref. 1). An information system compiles, communicates, and records all major decisions, actions, and developments. This information system is described in the quality manual. Changes due to the results of current research findings and problems play a particularly important role in this documentation and control system. Supervisors, funding agencies, and other authorized parties to the project require reliable and up-to-date information. Third parties and often the public want and need to understand what is being researched and the progress being made.

QA PROGRAM/SYSTEM

RATIONALE FOR QA PROGRAM

The advantages of a QA program for R&D are in the areas of product liability, government regulation, sales, and costs (Ref. 2). "Studies and experiments that are not conducted under controlled verifiable conditions and thereby produce data that are useless in supporting subsequent design decisions are a waste of time and money" according to George Hardigg, General Manager of Westinghouse's Advanced Power Systems Division (Ref. 2).

GETTING STARTED

In a well-managed research establishment and project quality assurance practices already exist, although they are mostly implicit. Starting at this point means to assess these practices and to translate them into formal procedures and documentation:

1. Collect data on existing systems and procedures.
2. Prepare process flow chart.
3. Prepare written descriptions of these basic systems.
4. Correlate the company's need and government regulation with current practices.
5. Modify the system as required.
6. Monitor the new system.

QA must have sufficient organizational freedom to initiate, require, and verify resolutions to quality problems, including "stop-work" authority (Ref. 2). Most of the actual assurance responsibilities are delegated to the operating project organization.

POLICIES AND PROCEDURES

QA starts at the top. Authority should be spelled out. The Quality Manual is often a series of policy statements. Topically it can be structured around generic standards (Ref. 2). Another possible layout is the following (Ref. 2):

1. Introduction
2. Management and Planning
3. Design and Development
4. Procurement
5. Manufacturing, Fabrication, and Assembly
6. Construction and Installation,
7. Operation, Maintenance, and Modification,
8. Quality Assurance Audits.

A similar content layout is (Ref. 4):

1. Planning and Organization
2. Qualification and Training
3. Control of Equipment and Material
4. Acquisition and Recording of Data
5. Peer Review
6. Audits

Implementing procedures should be written by those who are most affected by them, with QA reviews for consistency with policies.

QA should assess major projects; a form is displayed in Ref. 2, p. 17. Failure mode and effect analysis techniques are relevant here.

Standards for good laboratory practice and workmanship apply. Proprietary and classified material must be protected and publications should be vetted.

Project technical plans for individual experiments must be prepared to describe the purpose, activities, types of experiments, test equipment, materials, measurements, special processes, tests, data acquisition, analysis, and reporting. A special QA plan is available (Ref. 2, p. 37ff). QA verifies implementation of the technical plan. The benefit of written quality plans that meet recognized standards ensures that all research meet desirable criteria. The plan is simply an extension of accepted scientific ethics (Ref. 4). Many researchers find that these plans do not change their usual practices.

PROJECT DESIGN CONTROL

In a research situation design reviews for facilities follow normal rules and practices. For research projects these are best performed on a day-by-day basis with the focus on technical data. The general requirement for a nuclear design control program identifies design interfaces, coordination of participating design organizations, and translating of design basis into design documents.

The project test section design should be evaluated for (Ref. 2)

* conformance with project specifications,
* material compatibility,
* design interfaces related to fit of parts,
* dimensional stability,
* conformance with plant safety regulations,
* correctness of information used for calculations,
* verification of calculations,
* adequacy and correctness of computer programs,
* appropriateness of specified quality standards.

Proper documentation of all calculations—the objective, sources, assumptions, model/equations, software, results, and recordings—is of particular importance, as is the control of drawings by using standardized procedures.

The QA department verifies the effectiveness of the review system by audits.

PROCUREMENT

Critical items require special attention and review. The requisitioner determines the QA specifications. A QA review checks for general aspects and consistency. A qualified supplier list should be maintained. Other items to be checked include

* supplier understanding of specifications,
* compliance with standards,

* validity of certifications,
* calibration,
* supplier evaluation,
* source and receiving inspection,
* certification of conformance,
* certified test results.

MEASURING AND TEST EQUIPMENT

The validity and reliability of the research conclusions depend on the quality and maintenance of the equipment used. Many technical standards need to be applied, complied with, and recorded and documented. Traceability means that results can be traced to proper standards. With the new and constantly improving technology used for research, this is not an easy task. Measurement traceability is more feasible than that of standards traceability (Ref. 2). The program defines the scope; quality assurance levels; authorized checks; centralized controls; calibration procedures; and evaluation of agencies, logs, and history records.

TEST PLAN

This plan covers test planning, control, and use of equipment. The design of experiments is at the core of a test plan. Basic planning leads to the formulation of specific test procedures and steps and includes documentation and reviews.

SOFTWARE

Standards for software quality apply. For reliance on software and computer software, QA plans ascertain the adequacy of their design and application. The project's technical plan and the associated QA plan describe additional software quality control, quality levels, and instructions. Checklists are described in Ref. 2, as is a software development process (Ref. 2, p. 112).

RECORDS AND REPORTING

Careful logging and recording is an important responsibility for researchers and their technicians. Reports on research and its quality assurance demonstrate actual performance. The quality of the report format and content reflects on the quality of research itself. All major documentation and plans could be appended to the report or submitted at various stages of the project. The reader should be able to receive additional information from the underlying information system, such as tracing original data and plans.

Reports, records, and other documentation remain available long after the project is completed. Planning and controlling retention starts early in the project.

AUDITING THE PROGRAM

Reports on audits conducted during the project aid both management and researchers. These reports must be retained with the other project documentation.

Audits imply that a QA program and specific control procedures are in place, so that compliance and general effectiveness can be examined and evaluated.

Auditing standards provide guidance in the planning and execution of these special audits. They also describe the qualifications and status of auditors. Auditees, usually the researchers, will cooperate when the purpose, approach, and reporting have been clarified to the satisfaction of all participants.

Research projects demand that audit concepts and practices suit the situation, task, and personalities. There is considerable uncertainty and risk, rendering procedures and work standards subject to continuous interpretation, adaptation, and modification. An auditor proceeds with care, and remains in close contact with the auditee. Audits frequently initiate changes, or support researchers in bringing about desired and required improvements from their perspective.

Checklists, the main guide for the audit, can be made available along with the audit plan. Frequent meetings keep participants informed about progress and major findings. Once the facts are firmly established, corrective action can restore the desired state of control, which is to be verified by the auditor.

Internal audits often take the form of an independent "double check," possibly in preparation for a formal external audit. Researchers with assistance by QA staff can conduct their own audit. Traditional peer reviews, widely relied upon in the past, are not a substitute for these audits, although they are useful in their own right (Ref. 4).

The object of these audits needs to be clearly established and documented in the audit plan. Any area of the QA program, individual projects, processes, units, suppliers, contractors, etc., can be audited when planned or when a special development suggests such an audit. All areas should be audited at regular intervals as a matter of policy.

BIBLIOGRAPHY

1. Schimmel, G., Schramm, K. H.; Moeglichkeiten und Grenzen von Qualitaetssicherungssystemen in der angewandten Forschung; *Qualitaet und Zuverlaessigkeit,* vol. 32 (1987), No. 11, pp. 557–559 (in German).

2. Roberts, George W., *Quality Assurance in Research and Development,* New York, NY: Marcel Dekker, Inc., 1983.

3. Koning, John W., *The Scientist Looks At Research Management,* An AMA Management Briefing, AMA, COM, American Management Association.

4. Bussolini, P. L., Davis, A. H., Geoffrion R. R., A new Approach to Quality for National Research Labs, *Quality Progress,* Vol. XXI, No. 1 (January 1988), pp. 24–28.

5. Gushue, J. M., Shashidhara, N. S., QA for Hazardous Waste Management, *Quality Progress,* vol. XXI, No. 1 (January 1988) pp. 48–51.

Appendix A
Glossary

Acceptance Sampling Plan states sample size(s) and associated Acceptance Number (e.g., allowed defectives) for a given lot size.

Acceptable quality level (AQL) is the highest percentage of defectives, or proportion of variant units, allowed for good quality, and with high probability of acceptance.

Appraisal cost are costs for inspecting and testing products or services supplied, work in process, end-items, processes, returns, etc.

Audit (see under *Quality Assurance Audit*)

Audit department/unit programs audit projects, employs auditors, and administers the audits.

Auditee is the person, or unit, in charge of the object to be audited and sharing the responsibility for a successful audit.

Auditing standards are published documents that describe all important features of audits and auditing. These standards ensure the quality of audits and the resulting audit reports.

Auditor is the person, or team, qualified and appointed for an audit project. Auditors are often organized into an auditing unit. An auditor for a project is normally appointed by the senior management that initiated the audit and that will receive the audit report.

Corrective action is that determined and communicated by those persons or departments having the authority to make recommendations, such as inspectors or supervisors. Auditors usually do not have the authority. Corrective action should prevent the recurrence of problems.

Defect is any perceived fault, error, failure, or other fact or occurrence that bears negatively on the usefulness and expectation with regard to a product and service. Defect(s) can render an item or service defective. Such items are *defectives*.

Design is the result of researching customers' stipulations and specifications for products and services; determining product or service attributes and measurements; developing and field testing prototypes; determining the necessary technological and resource requirements; and documenting and finalizing the product or service.

Design assurance includes all decisions and activities directed toward establishing and meeting quality objectives to the customer's satisfaction. Thoroughly planned and documented quality specifications facilitate design quality control. It approves and documents attributes and measurements of a product or service.

Design effectiveness is the degree to which a design creates customer satisfaction within the limitations of price and other technological, social, and economic factors.

Drawing is any illustrative and schematic description of an item, or part thereof, for the purpose of communicating a design and facilitating its production. Drawing can be associated with explanatory documents.

End item is any product that is complete as defined by the design and is released for delivery by a production supervisor. For services it is the actual rendering of the service to the customer, e.g., the finalizing of an insurance contract.

Failure cost is caused by company or plant internal and external failures, such as avoidable scrap, rework, repair, returns.

Inspection is an independent verification of conformance to a standard (specification) using approved methods by a competent operator or inspector. Inspection can be manual or automated or computerized.

Inspection point is the production stage where the verification is to be performed according to certain approved procedures.

Lot tolerance percent defective (LTPD) is the level of poor quality with low probability of acceptance.

Nonconformance is any deviation in a product or service from specified quality characteristics. Inspection or test procedures normally define major nonconformance in order to alert the inspector. Nonconformance control should also include any perception of a deviation by other than official inspectors. "Nonconformance" is more narrow in concept than "defect."

Prevention costs are those resulting from the prevention of defects and failures, such as conducting audits.

Procedure is a formal and mandatory directive and guideline for work performance that provides all necessary information and performance criteria. Supervisory management plans, approves, and audits these procedures. Components of a Procedure are purpose, policy, application, references, definitions, performance steps and sequence, and performance criteria. A brief explanatory note may be added.

Process is any planned sequence of activities or operations aimed at a desired and specified outcome under specified conditions. The process is in a controlled condition when output or performance criteria are within tolerance or control limits.

Process capability study samples data from a process, determines the control chart criteria (centerline, upper and lower control limits), for normally $\pm$ 3 Standard Deviations and compares these criteria with the desired values. Deviations measure relative process capability and allow for adjustments, such as process improvement or change of desired values (tolerances).

Product includes material, parts, accessories, semifinished batches identified, and information brochures for the user.

Product assurance are all activities that plan and control quality-effective production and conformance to specification.

Productivity is the relation of output to input in a productive system with regard to the basic dimensions of material (quantity and quality), time, and place. Productivity improvement can be gained through continuous and systematic research and development, and the application of scientific knowledge and methods in all areas and by all people in an orgainzation. Projects are the most suitable vehicle for achieving productivity improvement that concur with higher quality.

Product status is a record or label identifying the product's condition for use, e.g., inspected, not inspected, in-transit, condemned, customer property.

Qualified supplier is one that has requested and undergone an inspection or audit by the potential purchaser or an agent of the purchaser, has been approved, and has been listed by the purchaser accordingly.

Quality assurance includes all activities to ascertain customer satisfaction with products delivered and services rendered. It includes planning and controlling quality.

Quality assurance audit or quality audit is a formal and systematic evaluation of the quality assurance system, especially compliance with applicable standards and its effective design and implementation. The object of the audit is predetermined and can be the entire quality assurance system, or individual procedures, product-based systems, contracts, or processes or plants.

Quality cost is the monetary value of resources expended for ensuring quality and for defective quality of products and services. It is the sum of prevention, appraisal, and failure costs. Both prevention and appraisal costs are quality assurance costs. Higher quality assurance costs should reduce failure cost—and in the long run quality costs. Other cost classifications amd compilations, such as fixed/variable, actual expenditure/opportunity cost, are applicable in quality cost accounting.

Quality Management System is all planned decisions and activities that jointly aim to attain the desired quality of products and services. It is also called a "Quality System." "Quality Control System," or "Quality Assurance System."

Quality plan is a document describing all product- or service-oriented activities including function, specification, point in the production process, test or inspection method, and decision rules. A quality plan when only describing inspection points and methods is called an "inspection plan" or "inspection checklist."

Receiving inspection is normally performed on the end item at the time of delivery and at the purchaser's plant or production plant or both. Inspection or test procedures are to be determined, communicated, and applied in cooperation with supplier and the quality assurance department.

Self-inspection is the operator inspecting his or her work in accordance with predetermined inspection or test plan and other applicable procedures.

Software encompasses all instructions and data inputs to a computer for electronic processing, and the associated procedures, documentations, manuals, and operating systems.

Software quality is the degree of intended fitness for use and actual user satisfaction. Quality characteristics are validity; reliability; ability to transfer, change, or adapt; ability to use the software for various applications; integrateability with other systems laterally and hierarchically; and ability to be standardized. The software should be free of errors and deficiencies; errors should be relatively easy to detect.

Software quality control is all decisions and activities to ensure software quality from the user's point of view. Planning precedes the actual control in that it sets performance specifications and standards. Individual quality controls are integrated and documented in a cohesively structured software quality control program. This program should be part of comprehensive quality assurance program or system.

Source inspection is a formal, systematic, and predetermined independent verification that product and process meet known and agreed-upon specifications. This verification is performed at the supplier's plant, or other external testing laboratory, with or without the presence of a purchase representative.

Specification means any quality characteristic, attribute, or measurement that is officially stipulated and approved. The totality of all product- or service-related specifications represents "quality" that is designed to meet customers' needs and requirements under given defined conditions for application and performance of the item, in a safe, reliable, and economic manner.

Statistical control chart shows control limits and a centerline. Plots from samples indicate the state of process control.

Statistical method means attaining information from numbers.

Statistical risks are those for rejecting a good lot (alpha risk) and for accepting a poor lot (beta risk).

Supplier inspection/audit is a formal and systematic evaluation of a supplier's ability to ensure quality to the satisfaction of a purchaser. The evaluation is commensurate with the relative need and importance of supply assurance, to be decided by the purchaser.

System standard documents the major characteristics of an adequate Quality Management System. The standard expresses the considered opinion of experts in the field and is verified through rigorous standard writing procedures.

Appendix B
Hospital Quality Assurance Manual (Outline)

This generic prototype manual is to help all those in the health care industry when they are preparing their own manuals. This outline will certainly need further additions and refinements.

The wide adoption of such a basic manual can help to comply with applicable standards for quality assurance programs. The still existing inconsistencies and unnecessary differences between generic and industry-related quality assurance program standards can be partly resolved through such a manual. Major concepts and methods of the standards are considered, and the user can then modify this manual to suit the needs and conditions of the health institution and organization.

The most important feature of a Quality Assurance Manual is its truthful description of all activities that are explicitly directed toward assurance of the quality of services and of the people rendering these.

QUALITY ASSURANCE PROGRAM MANUAL

for (name and address of institution)

Frontmatter

1. Title page
2. Preface (Foreword)
   General purpose of the Manual, overview of its main Sections.
3. Manual Writing
   Persons in-charge, initiation, terms-of-reference, writing procedure/project, drafting, changes, implementation, maintenance/audit
4. Acknowledgments

The Quality Assurance Program

PURPOSE

DEFINITION

For example, quality means the greatest achievable health benefits, with minimum unnecessary risk, and the use of resources in a manner satisfactory to the patient (Dept. of Health, Washington, D.C.).

THE INSTITUTION AND ITS MAIN MISSION

SCOPE AND APPLICATION

Delineate this Program from others; relate it as an integrated part of the institution and its operations.

The main program principles are related to program application, such as staff involvement in quality assurance and program.

APPLICABLE STANDARDS

References to all standards for this quality assurance program, including generic standards and excluding technical, non-quality-assurance standards.

PROGRAM ELEMENTS

1. The Program Objectives
2. Mission Statement
3. General Policy
4. Quality Assurance Commitment
5. Quality Assurance Program Principles
6. Organization
7. Institution

Quality Assurance
    a. Function/Department
    b. Program Coordinator/Manager
    c. Committees
    d. Job Descriptions

DEPARTMENT QUALITY ASSURANCE PROGRAM GUIDELINES

1. Principles
   Delegation of responsibilities, cooperation/coordination with QA coordinator, participation in QA program committee, reference to standards and guidelines.
2. Mission statement
3. Setting of goals of objectives
4. Intradepartmental QA organization
5. Staff relationships
   a. Service quality planning and control
   b. Performance standard
   c. Performance verification
   d. Performance improvements
6. Facilities maintenance
7. Setting of QA procedures and instructions
8. Individual care programs
   a. Responsibilities

b. Planning and control
   c. Quality Inspection
      Can include self-inspection, inspection instructions and reports, standardized and approved checklists
9. Quality Improvements
   a. Continuous training
   b. Improvement projects
10. Program Evaluation
    a. Internal audits
    b. Preparation for external audits
    c. External audits or reviews.

Departmental Quality Assurance

The scheme and guideline outlined should be adopted for each department. Departments should be allowed the necessary scope for meeting unique conditions. The involvement of a QA coordinator must be clarified and described.

Technical support for developing departments' programs is to be provided by the QA coordinator; this support includes standards and publications.

The program must be checked and approved at the institutional level.

Each program and form used must be fully documented. References to detailed instructions and technologies reduce the volume of the manual. The referenced documents must be readily available for audits.

Departments include:

1. Administration and Board

2. Medical Services

3. Nursing Service

4. Environmental Safety

5. Radiology Services

6. Emergency Services

7. Laboratory Services

8. Patients' Clinical Record Service

9. Pharmacy Service

10. Housekeeping

11. Rehabilitation Services

12. Respiratory Technology Services

13. Social Work Services

14. Therapeutic Dietary/Food Services

15. Long-Term Care

Emergency Plans and Environmental Services Quality Assurance Information System

This is a documentation of all reports and communication channels related to the Program. Prototypes and flowcharts simplify the overview and assessment.

QUALITY ASSURANCE PROGRAM MANUAL

RECORDS AND REPORTS

COMMUNICATION

RISK REPORTING, ALERT SYSTEM

QUALITY COST ACCOUNTING

STATISTICAL QUALITY CONTROL

STAFF QUALIFICATION AND PERFORMANCE

AUDITS AND REVIEWS

QUALITY IMPROVEMENT PROJECTS

PUBLIC RELATIONS

Appendix C
Quality Manual for a Small Manufacturer (Outline)

Quality Assurance Manual (Draft)

ABC Company

Policy Statement and Certification

Quality Assurance Policy

ABC Company and its staff, suppliers, and distributors are committed to ensuring the quality of its products and services in order to satisfy fully its customers. To this end a quality assurance program has been established that complies with Standard MIL-Q-9858A, *Quality Program Requirements*. Each Procedure is based on a special policy statement.

Certification

I hereby certify that this Quality Manual accurately and adequately describes the quality assurance system in use.

Signature: (Manager/Owner)

ABC COMPANY

Amendments

Page No.	Amendment No.	Date	Amendment Description	Approval

ABC COMPANY

Table of Contents

1.	General
1.1	Policy Statement and Certification
1.2	Amendments
1.3	Foreword
1.4	Organization
2.	Quality Assurance Procedures
2.1	Design Assurance and Contract Review
2.1.1	Design Assurance
2.1.2	Drawing and Specification Change Control
2.1.3	Design Review
2.2	Review of Contract with a Customer
2.3	Quality or Inspection Plan
2.4	Supply Assurance
2.4.1	Supplier Selection, Control, and Cooperation
2.4.2	Purchasing and Subcontracting Control
2.4.3	Source and Receiving Inspection
2.5	Production Assurance
2.5.1	Verification of Production Readiness
2.5.2	Work-in-Progress Inspection
2.5.3	End-Item/Service Inspection
2.5.4	Control of Nonconformance and Defect
2.6	Quality Control and Inspection Resources
2.7	Physical Product Integrity and Care
2.8	Customer Relation

2.9	Product Performance Related Services
3	Quality Management Information System
3.1	Quality Report and Reporting
3.2	Statistical Quality Control
3.2.1	Statistical Process Control
3.2.2	Acceptance Sampling
4	Quality Assurance Audit
	Appendix
A	Establishing Quality Assurance Procedures
B	Forms and Records (samples)

ABC COMPANY

Foreword

The Quality Assurance Program is designed, implemented, and frequently reviewed by the quality assurance department under the direct supervision of senior management.

1. Quality assurance is the assigned responsibility of every staff member, supplier, and company representative. Our policy is to prevent defects and other nonconformances.

2. This Quality Manual describes the Program and its individual quality assurance procedures. As a management program the major concern is the planning and control of all assurance decisions and activities that are to implement the general policy. Modern management concepts and methods have been applied and are applicable in this Program.

3. Our quality assurance program is to comply with major standards. Customers are invited to assess this Program against such standards. Their support for improving this Program for our mutual benefit is welcomed.

4. Required changes will be made by the quality assurance department and is noted in the Quality Manual. Holders of a copy of this manual should maintain an up-to-date copy.

ABC COMPANY

Organization

(An organization chart should be included at this point)

1. The quality assurance department reports directly to the senior manager and company president.

2. Each functional manager and supervisor have explicit quality assurance responsibilities in their respective job description. The quality assurance department provides support and guidance.

3. Individual staff members have an opportunity and incentive for recommending Program improvements and for actively participating in the quality assurance effort.

(The individual procedures completing the system for this company were adopted from those in this Guide and adjusted to meet the MIL 9858A Standard requirements. These procedures are not shown here.)

Index

Acceptable Quality Level (AQL), 113
acceptance sampling, 112
  procedure, 112, 113
  standard, 113
accounting and finance, 17
accounting, quality cost, 104
administration, 17
appraisal cost, 104
audit, 139
  audit checklist, 158
  checklist, 146
  example, 147
  design review, 55
  guideline, 139
  guideline/standard, 147
  hospital quality assurance, 167
  principles, 141
  procedure, 142
  research project, 190
  standard, 141, 142
  supplier, 62
  technical aid, 145
audit procedure, small business, 143
auditor, 141

Average Outgoing Quality Level (AOQL), 113
business, small, 5
  audit procedure, 143
  quality assurance system, 174
  quality manual (example), 203
  system design, 175
calibration, 92
cases, 161
  hospital quality assurance, 161
  research establishment quality assurance, 178
  small business quality assurance, 174
company functions, 7
contract
  control, 65
  review, 55
control chart, 109, 111
control chart criteria, 108
control limits, 108
corrective action, 70, 88
course, 6
customer, 97
customer relation, 96, 97
  audit checklist, 154

procedure, 96
customer service assurance, 93
defect, 88
defective, 88
design, effectiveness, 52
design assurance
  change control, 49
  design review, 52
  procedure, 45
design engineering, 15
design review, 52
  quality system audit, 55
end-item, inspection procedure, 85
failure cost, 105
general services, 17
glossary, 191
guide, 6
guideline
  audit, 139
  auditing, 147
  ISO, 21
  project management, 31
hospital
  quality assurance example, 170
  quality assurance manual, 197
human resources, 17
implementation, 124
  quality management system, 124
information system, quality
  management, 100
inspection, 78
  by operator (self-inspection), 80
  end-item, 85
  in-process, 77
  receiving, 70
  source, 70
inspection plan, 57
inspection point, 78
inspection resources, 92
  audit checklist, 153
ISO Guideline, 21
Lot Tolerance Percent Defective
  (LTPD), 113
marketing, 15
MIL-Std-105D, 113
MIL-Std-414, 113

nonconformance, 88
  report, 90
  review report, 90
nonconformance control
  audit checklist, 152
  procedure, 87, 89
prevention cost, 104
procedure
  acceptance sampling, 112, 113
  audit, 142
  audit (small business), 143
  customer relation, 96, 97
  design assurance, 45
  end-item inspection, 85, 86
  format, 44
  in-process inspection, 77
  inspection resource control, 91
  nonconformance control, 87, 89
  production start-up, 75
  prototype, 14
  quality assurance workshop, 133
  quality assurance workshop
    (simplified), 136
  quality cost accounting, 104, 105
  quality management system
    implementation, 126
  quality manual writing, 130
  quality plan, 58
  quality reporting, 101, 102
  self-inspection, 80, 81
  service quality assurance, 99
  software quality control, 115,
    117
  statistical process control, 109
  statistical process control
    (simplified), 110
  supply assurance, 62
  supply verification, 70
  workshop, 135
procedure writing
  projects, 40
  technical aid, 43
process, 78
process capability, 108
procurement, 16
product, 94

product integrity, audit checklist, 154
product performance, 93
product status, 94
product/service design assurance, audit checklist, 149
production, 16
  start-up, 74
production assurance, 74
  audit checklist, 151
production process, 7
production start-up, procedure, 75
project
  management, 31, 32
  procedure writing, 40
  quality improvement, 36
  research quality assurance, 183
project management, 33
quality assurance
  audit, 139
  developments, 3
  hospital system development, 169
  tasks, 14
  workshop, 133
quality assurance audit, 140
  audit checklist, 158
quality assurance program, 4
  auditing, 12
  design, 8
  paperwork, 13
  principles, 8
quality assurance survey, checklist, 28
quality assurance system
  action plan, 25, 26, 27
  audit, 139
  audit checklist, 149
  designing, 25
  hospital, 161
  research project, 186
  standards, 22
  subsystems, 45
quality assurance workshop, technical aid, 137
quality audit, 140
quality control, software, 115

quality control resources, audit checklist, 153
quality cost, 104
  accounting, 21
quality cost accounting, procedure, 104, 105
quality improvement
  projects, 36
  technical aid, 39
quality management, 3
  information system, 100
quality management information system, audit checklist, 154
quality management system
  implementation plan, 124
  implementation procedure, 126
  implementing, 123
quality manual, 10, 11, 127, 128
  audit checklist, 157
  format, 129, 130
  hospital (example), 197
  research project, 185
  small business (example), 203
  table of contents (example), 131
  technical aid, 132
  writing procedure, 130
quality plan, 57
  procedure, 58
quality report, 101
quality reporting, 101
  procedure, 101
quality system, 100
R&D project, quality assurance, 181
report, 102
  nonconformance, 90
reporting, 102
  procedure, 102
research
  quality, 181
  quality assurance, 181
research establishment, quality assurance, 178
research project, audit, 190
research quality assurance
  quality manual, 185
  technical aid, 182

resources
  inspection, 91
  quality control, 91
self-inspection
  mode, 84
  procedure, 80, 81
seminar, 6
service, 99
service quality assurance, 98
  procedure, 98
small business (see business, small)
software, 116
  quality control, 115
software quality, 116
software quality control, 116
  audit checklist, 156
  procedure, 117
  procedure (simplified), 120
  standard, 116
source inspection, 70
standard, 6
  acceptance sampling, 113
  audit, 142, 147
  determining, 23
  hospital quality assurance, 164
  quality assurance systems, 18, 19, 22
  statistical process control, 109
standard deviation, 108

statistical method, 108
statistical process control
  procedure, 107, 109
  simplified procedure, 110
statistical quality control, 107
  acceptance sampling, 112
  audit checklist, 155
statistical quality planning, 107
statistical risks, 113
supplier, qualified, 62
supply, verification, 69
supply assurance, 61
  audit checklist, 150
  procedure, 62
system implementation, audit checklist, 157
technical aid
  audit, 145
  procedure writing, 43
  project management, 34, 35
  quality assurance workshop, 137
  quality improvement, 39
  quality manual, 132
  research quality assurance, 182
workshop, 133
  audit checklist, 158
  principles, 134
  procedure, 133, 135